Genetic Sex Differentiation in Fish

Series on
Sex Differentiation in Fish

Genetic Sex Differentiation in Fish

T.J. Pandian
Visiting Professor
CAS Marine Biology
Annamalai University
Tamil Nadu
India

CRC Press
Taylor & Francis Group
Boca Raton London New York

CRC Press is an imprint of the
Taylor & Francis Group, an **informa** business
A SCIENCE PUBLISHERS BOOK

CRC Press
Taylor & Francis Group
6000 Broken Sound Parkway NW, Suite 300
Boca Raton, FL 33487-2742

Cover Illustration: For genetic therapy, stem cells have become a handy tool. But their availability is limited to the prenatal stage in mammals and other vertebrates. However, fishes are unique and generate them throughout their life time. The cover page shows the stem cells identified in Germinal cradles (arrows) of the ovary of *sox9b EGFP* transgenic medaka fish. The cradle comprises of stem cells and *sox9b* expressing cells, that are interconnected by cellular processes of *sox9b* expressing cells (arrow heads) to form the ovarian cord (from Nakamura et al. 2011. Ovarian germline sterm cells in the teleost fish medaka (*Oryzias latipes*). Int J Biol Sci, 7: 403–409, with kind permission of IJBS Publishing Team)

Printed in the United States of America on acid-free paper
Version Date: 20120327

International Standard Book Number: 978-1-57808-799-0 (Hardback)

Library of Congress Cataloging-in-Publication Data

Pandian, T. J.
Genetic sex differentiation in fish / T.J. Pandian. -- 1st ed.
p. cm.
Includes bibliographical references and index.
ISBN 978-1-57808-799-0 (hardback)
1. Fishes--Genetics. 2. Fishes--Genome mapping. 3. Fishes--Sexing. 4. Fishes--Reproduction. I. Title.
QL638.99.P466 2012
571.88217--dc23

2012009524

Visit the Taylor & Francis Web site at
http://www.taylorandfrancis.com

Science Publishers Web site at
http://www.scipub.net

CRC Press Web site at
http://www.crcpress.com

Preface to the Series

Fishes constitute more than 50% of the living vertebrate species. The level of sexual plasticity and flexibility displayed by them has no parallel among other vertebrates. With most divergent expression of sex including many morphants within a sex, sexuality in fishes ranges from gonochorism to unisexualism and to an array of unbelievable patterns of functional hermaphroditism; their mode of reproduction also ranges from sexual to gynogenesis and hybridogenesis, each of these modes involves external fertilization and oviparity or internal fertilization and viviparity, which itself ranges from zygoparity to embryoparity. In fact, no other vertebrate group offers as much diversity as fishes to study the processes of sex determination and sex differentiation.

This book series is a continuum of my books on 'Sexuality in Fishes' and 'Sex Determination in Fish'. Information available on Sex Differentiation in Fish is rich and diverse, academically and economically so important that the original plan of bringing out a single volume had to be changed to three separate books: 1. Genetic Differentiation, 2. Endocrine Differentiation and 3. Environment Induced Differentiation. An overview of a large number of publications concerning 'Sex Differentiation in Fish' suggests a transition from experimental to analytical approach during the last decade. With the advent of molecular biology and arrival of more sophisticated tools and techniques, publications in this hot area are covered by an increasing number of papers in a wide variety of journals. Consequently there is a wide gap between molecular biologists including molecular endocrinologists publishing in high profile journals and fisheries scientists who continue to publish in conventional fishery biology journals. A major objective of this book series is to bridge the widening gap between them. Nowadays single author book series has become a rarity but an attempt has been made to accomplish a comprehensive synthesis of available relevant publications in this area.

October, 2011 T.J. Pandian
Madurai

Preface

Sex is a luxury and costs time and energy but ensures recombination to generate genetic diversity. Fishes display a level of sexual diversity and plasticity that are unrivalled among other vertebrates. Consequently no other vertebrate group offers as much diversity for the study of sex determination and differentiation as fishes. One objective of this book is to bridge the widening gap between molecular biologists with publications in high profile journals and fishery biologists whose reporting is limited to conventional fishery biology journals.

This book, a comprehensive description of sex differentiation in gonochoric and hermaphroditic fishes, explores the differentiation process by examining it from sex chromosomes and sex determining genes through sex differentiation genes and sex steroid receptor genes to stem cells and germ cells. It (i) emphasizes the need to search for more than one sex determining gene residing on different chromosomes, (ii) evidences the brain sexualization of primary gonochores prior to sexual maturation from studies on transplantation of germ cells, endocrine sex reversal and gonadectomy, and for the first time indicates that gonadal differentiation implies reduction of bisexual potency of germ cells supporting somatic cells to unisexual potency, (iii) traces the origin of hermaphrodites from secondary gonochores, (iv) explains the conservation of homologous sex differentiation genes and steroid receptor genes among gonochores and hermaphrodites, (v) recognizes that the changes in the phenotypic morphotypes among gonochores is intrasexual but intersexual among hermaphrodites, (vi) identifies the optimal broodstock size, the appropriate candidate species for bait and food fishes and recognizes that the fishes do undergo menopause and reproductive senescence, which has an implication on the production of stem cells and (vii) shows that the importance of steriles and mutants for better understanding of the sex differentiation process of fishes, as model for human diseases and genetic therapy.

Genetic Sex Differentiation in Fish succeeds my earlier books Sexuality in Fishes and Sex Determination in Fish. This research area has moved from observational to experimental, and to analytical during the last decade. This book is a comprehensive synthesis of over 630 relevant publications collected from a widely scattered 210 and odd literature source. The citations of more than 77% of references in this book are dated after 2000, clearly indicating that this is a widely research area with active participation of about 3000 scientists around the world. This book is the first on this theme in fishes. Further two books are in the pipeline.

October, 2011 T.J. Pandian
Madurai

Acknowledgements

It is with great pleasure that I thank Dr E Vivekanandan and R. Koteeswaran, who have critically read the manuscript and offered many useful suggestions. I am grateful to Drs R Jeyabaskaran, G Kumaresan, P Murugesan and AG Ponniah for useful discussion. Special thanks are due to Prof T Balasubramanian for constant support and encouragement. I gratefully appreciate R Koteeswaran, who has very patiently searched relevant publications. I thank the Central Marine Fisheries Research Institute, Kochi, which has the best library in India for fishery science, for providing excellent literature service under the leadership of Mr V Edwin Joseph. The manuscript was prepared by Ms S Akila B.Sc., B.L and I wish to thank her for patience, cooperation and excellent work.

To reproduce figures and tables from published domain, I am to thank many. Firstly, I wish to thank my guru Prof O. Kinne for providing me copies of his journals. I wish to gratefully appreciate the open door policy and record my sincere thanks to The Genetic Society of America (*Genetics*), The American Society of Ichthyologists and Herpetologists (*Copeia*), The Brazilian College of Animal Reproduction (*Animal Reproduction*), The Society of Endocrinology and Bioscientifica (*Journal of Molecular Endocrinology*), The Fishery Society of the British Isles/John Wiley & Sons Ltd (*Journal of Fish Biology*), The Ichthyological Society of Japan (*Japenese Journal of Ichthyology, Ichthyological Research*), The Japanese Society of Fishery Science (*Fishery Science*) and The Zoological Society of Japan (*Zoological Science*). I am equally thankful to the editors/Publishers of the following journals: *Aquatic Biology, Biocell, Biological Bulletin, Bulletin of Nansei National Fisheries Research Institute,* Cambridge Universiy Press, *Current Science, Development Growth and Differentiation, Fisheries Bulletin, International Journal of Biological Science, International Journal of Developmental Biology, Journal of Faculty of Applied Biological Science* and *Proceedings of National Academy of Science, USA.* I welcome and gratefully appreciate the new policy of, *European Journal of Neuroscience, Journal of Fish Biology* and *Reproductive Biology and Endocrinology* for choosing to leave the copyright with respective authors, and the open access policy of the *BMC* and *PLoS* Journals. Accordingly I thank very sincerely my fellow scientists: Drs LR Franca, A Jorgensen,

O Kah, S Kishi, Y Kobayashi, D Reznick, M Schartl, KR Siegfried and CA Strussmann. Dr Strussmann also facilitated the permission securing process with a couple of Japanese publishers. I also record my sincere thanks to *Development* for issuing permission with no charge. Special thanks are also due to Drs A Balasubramani, K Coward and NR Bromage, JP de Bruin and his co-authors, C Minto and CP Nolan, whose data have been subjected to recalculation to arrive at new ideas. For advancing our knowledge in this area through their rich contributions, I thank all my fellow scientists, whose publications are cited in this book.

T.J. Pandian

Contents

1

Introduction

Fishes are a fascinating group of animals, known for their sexual diversity and plasticity. Their indecisive mechanism of sex determination and flexibility in sex differentiation are the subject of numerous studies and publications. Sex determination is an event, may be a special one, that lasts for minutes and seconds, but sex differentiation that follows sex determination is a life long process. They are intimately interconnected and many biologists have treated them together (e.g., Devlin and Nagahama, 2002; Morinaga et al., 2004), some have even referred one for the other. With the advent of molecular biology, this area of sex determination and differentiation has become a 'hot spot', for molecular biologists, who are in search of sex determining genes and publish in high profile journals, and fishery biologists, who are interested in monosex aquaculture and are hindered with conventional fishery biology journals. One objective of this book is to bridge the widening gap between them.

1.1 Sex determination

It is that opportune moment of probability of sperm-egg encounter, which determines the sex of a progeny arising from the fertilized egg. Subsequently, the sex differentiation process, through which the progeny is sexualized, is initiated and may last almost throughout the life span. In fishes, the sex determination process itself is not decisive, as it is in mammals, and the sex differentiation process in fishes is even more indecisive and is influenced by an array of environmental factors like density, hypoxia, pH, temperature etc. (Pandian et al., 2012). Further the process is also amenable to sex reversal by induction of ploidy (Pandian, 2011), and a host of endocrine and other chemicals (Pandian, 2012).

Despite having diverged from the tetrapod lineage some 450 million years ago (Mya), the teleost fishes still share 70% of the genomes with other vertebrates and many central mechanisms underlying reproduction (Le Page et al., 2010). The root causes for the observed plasticity in sex determination and differentiation processes of fishes have been traced to two unique characteristics: 1. The non-sexualization of fish brain permanently, unlike in mammals and 2. The capacity of fish brains to constantly grow during adulthood and to regenerate after injury, and high aromatase activity (Table 1). Consequently, the critical window of brain sexualization, which is limited to the prenatal period in mammals, is extended through the entire life span of fishes (Fig. 1). However, this generalization of Le Page may be true of secondary gonochoric and hermaphroditic fishes; the critical window of brain sexualization seems to be closed in most primary gonochoric fishes at the juvenile stage or latest by puberty.

In a classical contribution, Le Page et al. (2010) have summarized the major differences between mammals and fishes in gonadal differentiation and brain sexualization (Table 1). The notables are: (i) sex is decisively determined by a single *Sry* gene located on a morphologically distinguishable sex chromosome in mammals but it may be by many genes (e.g., Hale et al., 2010) located on different and not readily distinguishable chromosomes in fishes (e.g., Bradley et al., 2011), barring *Dmrt1bY* in a couple of medaka fishes (Matsuda et al., 2002; Nanda et al., 2002), (ii) restricted to a small area, the brain is sexualized during the prenatal period in mammals but the sexualization continues in the entire brain throughout the development and adult life in fishes, (iii) testosterone is the male hormone in mammals, but it is 11-Ketotestosterone (11-KT) in fishes and (iv) there is only one aromatase gene with high expression in the ovary and associated functions in mammals; however, there are two aromatase genes *cyp19a1a* in the gonad and *cyp19a1b* in the brain of fishes (Le Page et al., 2010). Whole genome duplication inclusive of sex steroid receptor genes, and subsequent loss of (androgen receptor) *ar* A and substitutions of *ar* B are viewed as a permissive factor allowing plasticity and evolvability of divergent sex determination mechanisms in fishes (Douard et al., 2008).

Of 30,000 and odd fish species, cytogenetic and genetic analyses have so far been made only for 1,700 species. The presence of sex chromosome and heterogamety has been recognized only in 11% of them. Following many efforts to search for a series of sex linked genes, it has become possible to locate the sex determining loci, that is, the SEX in *Salmo salar* (Artieri et al., 2006). Clearly, it may not bear much fruit to endeavor to identify a single sex determining gene located on a sex chromosome in fishes. Under the influence of the mammalian scenario, it appears that many fish geneticists and molecular biologists are searching for a single sex determining gene residing in a specific sex chromosome. As indicated, the piscine genetic mechanism of sex determination seems to reside in more than one gene

Table 1. Main characteristics and differences between mammals and teleost fishes regarding gonadal differentiation (from Mouriec et al. 2009; Le Page et al., 2010; condensed)

Mammals	Teleosts
1. Sex determination	
Genetic determination only	Genetic determination only in a couple of medaka
Sry is the only master male determining gene	Many others, temperature, hypoxia, density, social environment also determine sex
2. Neurogenic activity	
High during development only	High throughout development and adult life
3. Hormones	
(i) Testosterone	
Crucial for brain masculinization, Stimulates secondary sexual characters	Not a male specific hormone Involved in final maturation of ♂ and ♀
(ii) Estrogen	
Required for brain feminization via testosterone aromatization, regulation of ovarian cycle	Triggering and maintenance of ovarian differentiation, stimulates vitellogenesis, strongly up-regulates brain aromatase
(iii) 11-Ketotestosterone	
Not present	Key androgen hormone in most fishes, involved in testis differentiation
(iv) 11β Hydroxy testosterone	
Not present	Key androgen hormone in some fishes, involved in testis differentiation, stimulates secondary sexual characteristics, modulates adult sexual behaviour
(v) Aromatase	
One aromatase gene but different expression regulated at promoter level. Low expression in testis, high expression in ovary and associated functions	Gonad specific gene *cyp19a1a*. Down- regulated during testis differentiation, up-regulated during ovary and associated functions
(vi) Aromatase in brain	
Key role in brain sexualization during prenatal period but moderate expression in adult. Mainly expressed in neurons. Radial glial cells disappear following embryonic neurogenesis	Extraordinarily high levels (100 to 1000 fold) brain specific *cyp19a1b*. Involved in brain sexualization throughout life. *cyp19a1b* activity restricted to radial glial cells from forebrain. Low expression on embryos but high in adults of both sexes

located on different chromosomes. For instance, a QTL sex determining gene is linked to another QTL for body size in *Sparus aurata* (Loukovitis et al., 2011) and *Dmrt1* is located on chromosome 5 and *cyp21a2* on chromosome 16 in *Danio rerio* (Bradley et al., 2011). Hence this presentation emphasizes

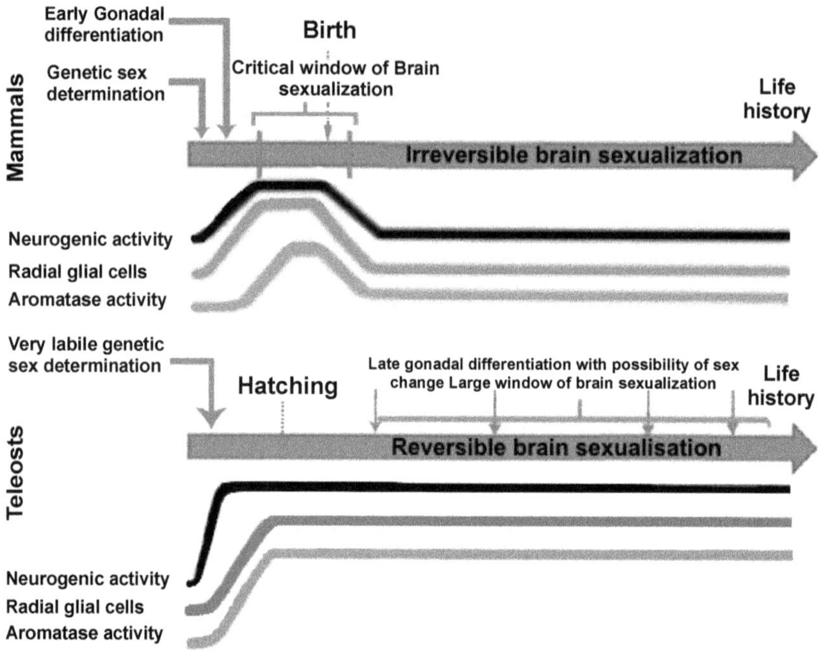

Fig. 1. Ontogenetic pathways of sex determination and differentiation in mammals and fishes. Note the differences in the critical window of brain sexualization (adopted from Le Page et al., 2010)

Color image of this figure appears in the color plate section at the end of the book.

the need for genome-wide linkage analysis and search for more than one sex determining gene located on different chromosomes of fishes.

In the context of sex determination in fishes, the hermaphrodites are of special interest. The hitherto cytogenetic search for the sex chromosomes in a dozen or so hermaphroditic fishes has identified the presence of an XX/XO system in the protogynic sequential labrid *Coris julis* only (see Devlin and Nagahama, 2002). It remains to be seen whether *C. julis* is a somatic XX hermaphrodite producing XX female and XO male progenies due to non-disjunction of one X chromosome in an embryo and an adult female resulting in the production of primary and secondary males, respectively (see Zarkower, 2006).

1.2 Hermaphroditism

In gonochores the male and female reproductive function is discretely expressed in *different individuals* but *within a single individual* in hermaphrodites. In fishes, simultaneous, sequential and serial are the recognized patterns of hermaphroditism (Fig. 2). The sequentials and

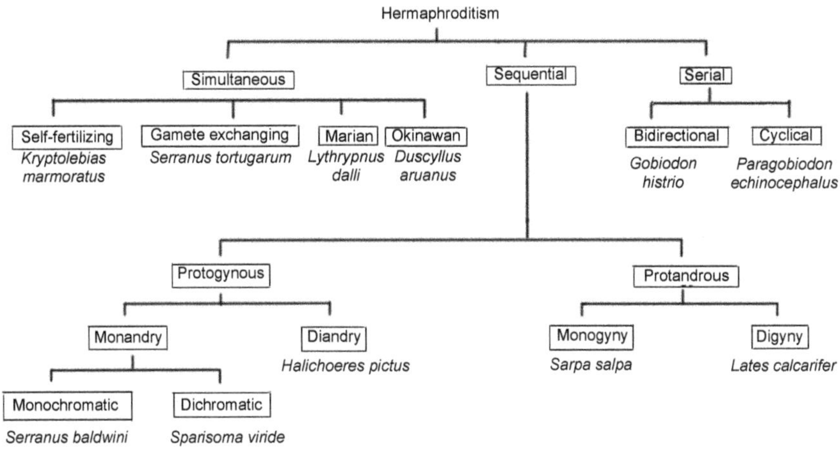

Fig. 2. Patterns of functional hermaphroditism in fishes (from Pandian 2011, modified)

serials change sex naturally once in the former but many times in the latter. The sequentials are further divided into (i) female to male sex changing protogynics and (ii) male to female sex changing protandrics. In the monandric protogynics, secondary males arise indirectly from sex changing females but in the diandrics primary males also arise directly, in addition to secondary males arising from sex changing females. For detailed information on sex-change mechanisms, Ross (1990) may be consulted. Marian and Okinawan hermaphrodites simultaneously possess the ovarian and testicular tissues in their gonads but essentially function as male or female and may change sex rarely (e.g., Asoh, 2003; Pandian, 2010). A dozen or so self-fertilizing and gamete exchanging simultaneous hermaphrodites usually do not change sex.

Hermaphroditism is polyphyletic and derived from rudimentary or juvenile hermaphroditic secondary gonochores. Documented in about 2% of teleosts, the presence of hermaphroditism is scattered across 20 taxonomic families in nine orders (Avise and Mank, 2009). Simultaneous hermaphroditism is reported from Cyprinidontiformes, Perciformes, Auloformes and Anguilliformes; protandry is known from Perciformes, Stoniformes, Siluriformes and Clupeiformes, while the protogynics from Cyprinidontiformes, Synbranchiformes, Perciformes and Anguilliforms (Sadovy de Mitcheson and Liu, 2008).

1.3 Patterns of sex differentiation

In fishes, sex differentiation is a more labile and flexible process, subjected to natural and artificial induction of sex reversal through environmental (Pandian et al., 2012), endocrine (Pandian, 2012) and ploidy (Pandian, 2011) manipulations.

Table 2 lists the lability of the sex differentiation process in selected fishes, in which sex reversal is/can be induced. The magnitudes of the lability clearly show that the early stages of fishes have retained bisexual potential to differentiate either into male or female. However, it is difficult to reconcile with differentiation in sex changing hermaphrodites. Incidentally, the term sex change refers to the natural event of gender change in a fish and sex reversal to artificially induced gender change. The wrong usage of terms like 'natural sex inversion' (Kokokris et al., 2006), 'sex inversion' (Li et al., 2006) and involution (Liarte et al., 2007) may be avoided.

Table 2. Lability of sex differentiation of selected fishes; the lability provides scope for (1) ploidy, (2) endocrine, (3) thermal or (4) environmental (4a: pH, hypoxia, 4b: density or 4c: surgical) manipulations. +* = all female brood, +** = all male brood, + = experimentally proved amenable, – = experimentally proved not amenable or not known (from Pandian and Koteeswaran, 1999, updated)

Family	Species	1	2	3	4a	4b	4c
Siluridae	*Clarias gariepinus*	+	+	–	–	–	–
	C. lazera	–	+	+	–	–	–
	Ictalurus punctatus	–	+	+	–	–	–
Anabantidae	*Betta splendens*	+	+	–	–	+	+
	Trichogaster trichopterus	–	–	–	–	–	+
	Macropodus opercularis	+	+	–	–	+	–
Cichlidae	*Oreochromis mossambicus*	+	+	+*	+	–	–
	O. aureus	+	+	+**	–	–	–
	O. niloticus	+	+	+	+	–	+
	Tilapia macrocephalus	–	–	–	–	–	–
Salmonidae	*Oncorhynchus mykiss*	+	+	–	–	–	+
	O. masou	+	+	–	–	–	–
Poecilidae	*Poeciliopsis lucida*	–	–	+	–	–	–
	Poecilia reticulata	–	+	–	–	–	–
Cyprinidae	*Cyprinus carpio*	+	+	–	–	–	–
	Danio rerio	+	–	–	+	–	–
Adrianichthyidae	*Oryzias latipes*	+	+	+*	–	–	+
Cobitidae	*Misgurnus anguillicaudatus*	+	–	+*	–	–	–
Paralichthyidae	*Paralichthys olivaceus*	+	+	+*/+	–	–	–
Atherinidae	*Odontesthes bonariensis*	+	+	+	–	–	–
	Patagonia hatcheri	–	–	+	–	–	–

Gonochores: Based on the presence or absence of a non functional intersexual phase during ontogenetic pathway of sexual differentiation, Yamamoto (1969) recognized two categories of sexual differentiation patterns and named them as the primary and secondary gonochores. Primary or differentiated gonochorism exists, when the differentiation process proceeds according to genetic sex determination. Hence the dimorphic sexual differentiation becomes recognizable prior to hatching, as in the Japanese medaka fish *Oryzias latipes* or after hatching, as in the Nile tilapia *Oreochromis niloticus* (Fig. 3; see also Fig. 14). *Oncorhynchus kisutch, Abramis abramis, Esox masquionongy, Sebastes schlegeli, Dicentrarchus labrax* and *Cyprinus carpio* are all good examples for the primary gonochores (see Devlin and Nagahama, 2002). A variant of the primary gonochores is the tertiary gonochore in which the bipotential gonad develops directly into either an ovary or testis (e.g., *Anguilla anguilla,* Beulbens et al., 1997, however see also Grandi and Colomba, 1997). Incidentally, the primary or differentiated gonochores commence with undifferentiated gonad, whereas undifferentiated or secondary gonochores with gonads that are already differentiated but non-functional as ovary/ovotestis (see Fig. 3). Hence the expression of 'differentiated' and 'undifferentiated' gonochores is a misnomer; they may preferably be named as primary and secondary gonochores.

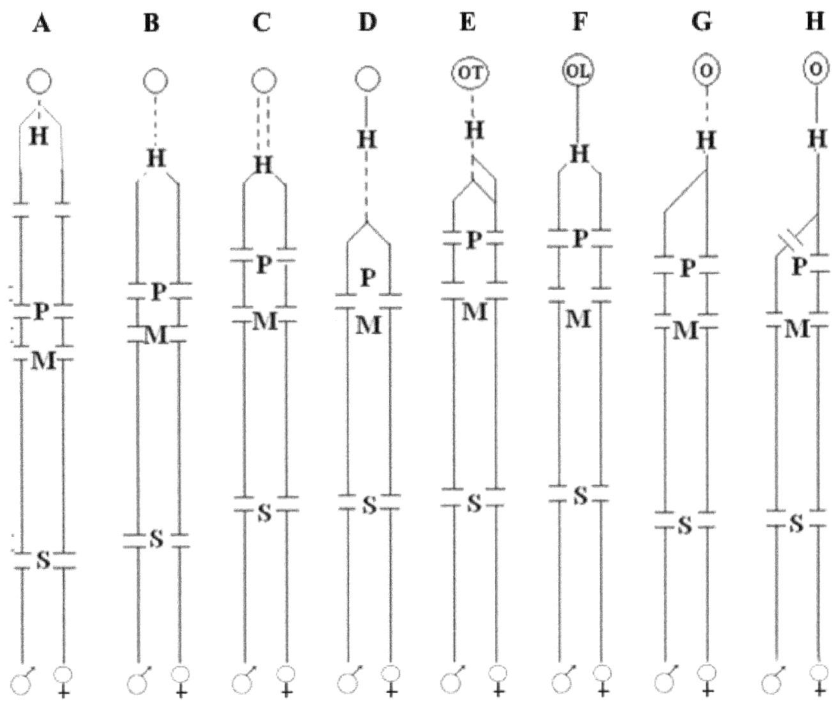

Fig. 3. Ontogenetic pathways of sex differentiation in selected gonochores. Primary gonochores : A = *Oryzias latipes,* B = *Oreochromis niloticus,* C = *Anguilla anguilla,* Secondary gonochores: D = *Pterogymnus laniarius,* E = *Mycteroperca rosacea,* F = *Pseudorasbora parva,* G = *Danio rerio* and H = *Dascyllus flavicaudus.* OT = Ovotestis, OL = Ovarian lumen, H = Hatching, P = Puberty, M = Maturation, S = Senescence?

On the other hand, secondary or undifferentiated gonochorism exists, when genotypic and phenotypic sexes differentiate in parallel lines for a short or longer duration but only one sex, as genetically or environmentally determined, becomes functional and the other becomes latent with persistent useful (e.g., *Nemipterus bathybius*, Takahashi et al., 1989) or vestigial tissue, or may not persist. Intersexuality is the presence of both male and female characteristics or of intermediate sexuality in a single individual. All simultaneous hermaphrodites are intersexes but some intersexes are not simultaneous hermaphrodites. The essential difference is the mutual tolerance between the two opposing sexual tendencies in simultaneous hermaphroditism. On the other hand, intersexuality is the result of interference between the two opposing sexual tendencies; consequently, neither of them can express themselves decisively and remain non-functional. Notably, the ontogenetic pathway of sex differentiation is interrupted by a duration of non-functional ovarian or intersexual phase, which may be better named as rudimentary hermaphroditism, or juvenile hermaphroditism. The term rudimentary hermaphroditism is defined as the existence of non-functional hermaphroditism *prior to puberty* and that of intersexuality to the transitional stage existing in a sex changing *mature adult*. Juvenile hermaphroditism refers to the existence of all juvenile females with ovary and its cavity, from which presumptive males arise by the transformation of the ovary into functional testis. These restrictions and definitions have become necessary, as authors indiscrimately and confusingly use various terms such as 'juvenile intersexuality', 'juvenile hermaphroditism', 'rudimentary hermaphroditism', 'non-functional intersex', 'juvenile bisexuality' and so forth (see bold lines in Tables 3, 4, 5).

Depending on the duration of the existence of hermaphroditism and others, secondary gonochorism is divided into four types:

Type 1. *Somatic* (intersexuality) *hermaphroditism* (Takahasi and Maeno, 1986), in which the presence of ovarian lumen is recognizable in all individuals (Fig. 3) but it may only be persistent in others, and may even be useful as a testicular cavity, formed in a similar manner as the ovarian cavity, as in *Pseudorasbora parva* (Table 3). However, the cavity is not detectable in others such as *Carassius auratus* and *Rhodeus ocellatus ocellatus* due to the overgrowth of testis. Some germ cells in differentiating testes undergo oogenesis in many cases.

Type 2. *Late gonochorism*, in which the duration of the rudimentary hermaphroditism is not transient as in somatic hermaphroditism, but lasts upto sub-adult or puberty stage of the fish (Fig. 3). With their delimited type of gonad, the sparid panga *Pterogymnus laniarius* and bronze bream *Pachymetopon grande* are good examples (Table 3). The hermaphroditism

Table 3. Ontogenetic pathways of sex differentiation in different types of secondary gonochores

Species/References	Reported observations
Type 1	
Pseudorasbora parva Takahashi and Maeno (1986)	**Juvenile hermaphroditism** is rather obscure. However, a testicular cavity, homologous to ovarian cavity, is present. Oogenesis within testis
Carassius auratus, Rhodeus ocellatus ocellatus (see Takahashi and Maeno, 1986)	Vestigeal ovarian cavity becomes not detectable, as testis grows over it
Nemipterus bathybius, N. virgatus Takahashi et al. (1989)	***Rudimentary hermaphroditism*** Ovarian cavity persists as vestige in testis. Functional testis substantially ovarian. Delimited gonad
Type 2	
Pachymetopon grande Buxton and Clarke (1992)	***Rudimentary hermaphroditism.*** Non functional dominant ovary and narrowly ridged testis. Delayed differentiation into functional ovary and testis. Delimited gonad
Pterogymnus laniarius Booth and Buxton (1997)	***Non functional intersex.*** Dominant ovary and posterolaterally ridged testis. Delimited gonad from which ovary and testis arise rather lately
Type 3	
Elacatinus evelynae, E. illecebrosus Cole (2008)	Gobiids. Transient ontogenetic expression of bisexual gonads prior to differentiation into males and females
Mycteroperca rosacea Erisman et al. (2008)	Commence as immature bisexuals. Some directly differentiate into females, while others remain immature bisexuals, from which females and males arise. Sex ratio 1♀ : 1♂
Epinephelus striatus Sadovy and Colin (1995)	Commence as females and bisexuals. Females arise directly as well as from bisexuals. Males arise only from bisexuals. Dygynic? Sex ratio changes from 0.71♀ : 0.29♂ to 0.5♀ : 0.5♂ with increasing body size from 42 to 52 cm
Type 4	
Barbus tetrazona tetrazona Takahasi and Shimizu (1983)	***Juvenile intersexuality.*** All gonads initially with ovary. Intersexual gonads degenerate in presumptive males between 45 and 50 dph. Ovarian cavity persistent. Sex ratio 1♀ : 1♂
Danio rerio Uchida et al. (2002)	***Juvenile hermaphroditism.*** All gonads initially with ovary. Intersexual gonads degenesate by apoptosis in presumptive males between 15 and 27 dph. Ovarian cavity persistent. Sex ratio ≈ 1♀ : 1♂
Dascyllus albisella, D. flavicaudus, D. trimaculatus Asoh et al. (2001), Asoh (2004), Asoh and Kasuya (2002)	Undifferentiated gonads differentiate into ovaries with oocytes developing upto cortical alveolar stage. Now many ovaries develop into matured ones while others into testis, following degeneration of oocytes but with persistent ovarian lumen

Table 4. Ontogenetic pathways of sex differentiation in monandric protogynic and protandric hermaphrodites

Species/References	Reported observations
Protogynics : Serranidae	
Cromileptes altivelis Liu and Sadovy de Mitcheson (2009)	Undifferentiated gonadal primordium develops an ovarian lumen. Meiotic oocytes are then formed. Differentiate into functional ovary and testis. A bisexual phase is missing
Sparidae	
Pagrus major Matsuyama et al. (1988)	**Juvenile bisexuality**. Gonad with ovarian lumen. Testis and ovary arise from bisexual gonad. Delimited gonad. Sex ratio 1♀ : 1♂. Transitional 13–33%
Gobiidae	
Tigrigobius multifasciatum Cole (2008)	Commences with ovary. Differentiates into transient expression of bisexual gonad. Testis and ovary are differentiated from the bisexual gonad
Labridae	
Labroides dimidiatus Kobayashi (1999)	Early gonadogenesis. Differentiation into intersex gonad. Some of them differentiate into testes, while others into ovaries
Parapercidae	
Parapercis snyderi Kobayashi et al. (1993)	Early undifferentiated gonad become intersexual. All of them differentiate into ovaries. Secondary males arise from females
Protandrics : Sparidae	
Diplodus sargus (Coetzee, 1986) *D. vulgaris* (Pajuelo et al. 2006)	Males and females are seen to arise from special types of protandric juvenile hermaphrodites
Sarpa salpa Joubert (1981)	***Juvenile hermaphroditism***. All mature males retain the vestigial juvenile ovarian tissue

is prolonged for 5.5 years, when the bisexual gonad sexually matures and differentiates into male and female bronze bream.

Type 3. *Rudimentary hermaphroditism* is observed among the gonochores of Serranidae, a family known for simultaneous and sequential hermaphroditism. In this type, all of them commence as rudimentary hermaphrodites; some directly differentiate into females (Fig. 3), while from the remaining both males and females arise. *Mycteroperca rosacea* is a good example (Table 3).

Type 4. *Juvenile hermaphroditism* is a well known type as documented in *Danio rerio*. In this type, all gonads first develop into ovary with oocytes and ovarian cavity (Fig. 3). Later, the oocytes are degenerated by apoptosis in half of them and the degenerated gonad in the presumptive males differentiate into normal testes. The transformation from the ovary to

Table 5. Ontogenetic pathways of sex differentiation in diandric protogynic hermaphrodites

Species/References	Reported observations
Labridae	
Halichoeres poecilopterus Kobayashi and Suzuki (1994)	Early gonads with many oocytes. Differentiates into intersex gonad; of the 14 intersexes, 6 differentiate into testis of primary males and others into functional females, from which secondary males subsequently arise
Suezichthys gracilis Kobayashi (1999)	Early ovarian gonad. Differentiate into intersex gonad. Half of them differentiate into testis of primary males and others into ovary of females, from which secondary males arise
Sparidae	
Pagrus pagrus Kokokiris et al. (1999)	***Intersexuals*** with ovary or undifferentiated gonad. Differentiates into ovary and testis of primary male. From females, secondary males arise. Delimited gonad. Sex ratio 1♀ : 1♂. Transitional 4.4%
Boops boops Monteiro et al. (2006)	Males and females are suggested to arise from ***juvenile hermaphrodites***. Majority of males originate early but some secondary males from sex changed females. Sex ratio 0.6♀ : 0.4♂. Transitional 5–18%
Serranidae	
Epinephelus coioides Liu and Sadovy de Mitcheson (2009)	Undifferentiated gonadal primordium develops ovarian lumen. Meiotic oocytes are then formed. Differentiates into ovary, and subsequently to bisexual phase, the ***Juvenile hermaphroditism***. Differentiates into functional ovary and testis in primary male. Secondary males arise from females
E. akaara Tanaka et al. (1990)	Commence with immature ovaries. Differentiates into bisexual gonads, subsequently into ovaries and testes of primary males. Females change to secondary males

testis occurs during the early juvenile stage, as in *D. rerio* or at puberty, as in *Dascyllus flavicaudus* (Table 3). The type 4 juvenile hermaphroditism represents the ancestors, from which the protogynics have been derived. The genus *Dascyllus* comprises of 10 species; *D. flavicaudus, D. albisella* and *D. trimaculatus* represent juvenile hermaphroditism, while *D. reticulatus* and *D. marginatus* are protogynics and *D. aruanas* is an Okinawan hermaphrodite. Thus the damselfish *Dascyllus* provides an excellent system for a study of the evolution of protogyny (McCafferty et al., 2002). Likewise, type 3 secondary gonochores seem to be the ancestors, from which the diandrics may have been derived.

Hermaphrodites: Gonochorism and simultaneous hermaphroditism are evolutionary end points on a reproductive spectrum (Avise and Mank, 2009). Incidentally, the hermaphroditic nematode *Caenorhabditis elegans* is a somatic XX female, in which the loss of an X chromosome due to nondisjunction results in the production of 5% XO males (Zarkower, 2006). It remains to be seen whether the self-fertilizing hermaphrodite *Kryptolebias marmoratus* and gamete exchanging hermaphrodites *Serranus baldwini* and *S. fasciatus* are also somatic XX females and whether the loss of an X chromosome due to non-disjunction produces XO males, induced by exposure to different temperatures in *Kryptolebias* and reproductive senescence in *Serranus.*

Like the secondary gonochores, all sex changing hermaphrodites commence with gonads that are already differentiated into ovary/ovotestis (see Figs. 3 and 4). Hence hermaphroditism has been derived from secondary gonochores possessing non-functional bisexual gonad. Most monandric sex changers pass through a common pathway (Table 4). Typically, the labrids, gobiids and parapercids, possessing undelimited type of gonad commence with bisexual gonads; on differentiation into ovaries, functional females are developed. Concomitant with sex change, the ovaries degenerate and functional testes are derived from the former ovarian wall (see Fig. 53) and the secondary males are produced (Fig. 4). However, the sparids possess delimited type of gonad and commence with ovaries, which subsequently differentiate into the bisexual gonad. In the serranid *Cromileptes altivelis* alone, the gonadal primordium, following the formation of ovarian cavity, directly differentiates into ovary and testis. The situation is similar to that of somatic hermaphroditism, in which the appearance of ovarian lumen is formed prior to the differentiation of the ovary and testis.

There is a uniform and common ontogenetic sex differentiation pathway for the diandric protogynic sex changers. In all of them, undifferentiated gonad and gonad with ovarian lumen (e.g., *Pagrus pagrus*) or ovarian gonad is formed the first (Table 5). However, they all differentiate into bisexual gonads. On further differentiation, ovaries and testis of primary males are formed. With the onset of sex change, the ovaries in female are degenerated and the testes of secondary male are derived from the ovarian wall of the female. There is paucity of information on sex differentiation of protandric sex changers. They seem to be derived from a special type of juvenile hermaphrodites (Fig. 4).

1.4 Impact of new techniques

This book is a continuum of the earlier books in this series. The first one Sexuality in Fishes is based on 460 or so publications from 175 literature sources that describe mostly observations supported by histological studies with occasional field or laboratory experimental observations on the succession of harem master by an α-female or harem mistress by an α-male. Cytogenetic contributions from over 600 publications that are scattered in 175 and odd literature sources formed the backbone of the second book 'Sex Determination in Fish'. In this book, relevant information from 620 publications that are very widely scattered in more than 210 literature

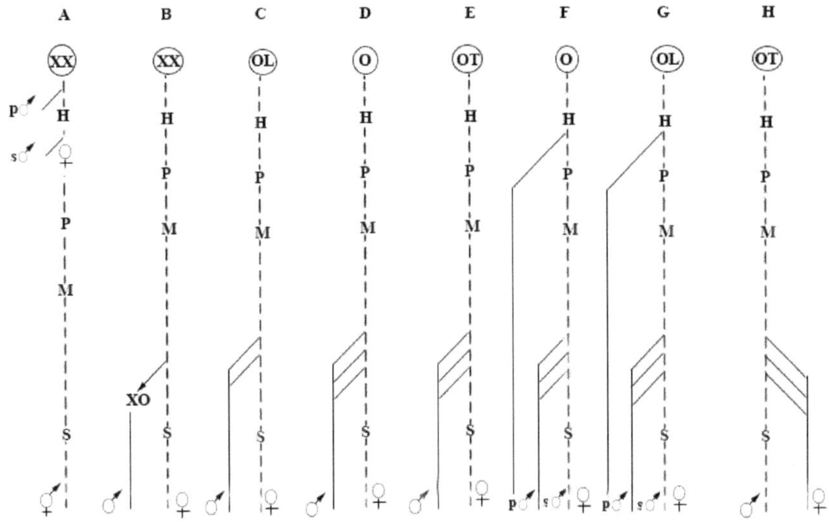

Fig. 4. Ontogenetic pathways of sex differentiation in selected hermaphrodites. Simultaneous hermaphrodites: A = *Kryptolebias marmoratus*, B = *Serranus baldwini*, Monoandrics: C = *Cromileptes altivelis*, D = *Tigrigobius multifasciatum*, Diandrics : E = *Parapercis snyderi*, F = *Suezichthys gracilis*, G = *Pagrus pagrus*, Protandric : H = *Diplodus vulgaris*. XX = Somatic egg, O = Egg, OT = Ovotestis, OL = Ovarian luman, H = Hatching, P = Puberty, M = Maturation, S = Senescence? p♂ = Primary male, s♂ = Secondary male, XO = Male from sex changed hermaphrodite

sources is synthesized. These publications elucidate the process of Genetic Sex Differentiation in Fish employing sophisticated modern techniques capable of very precise analyses. Incidentally, 50–52% of the references cited are dated after 1995 in the first two books but more than 77% are dated after 2000 in this book. Hence the publications on sexuality, sex determination and differentiation in fishes have projected the research into a 'hot area' and are being covered by increasing numbers and kinds of journals; for instance, from the Journal of Morphology and Animal Behaviour to Cell, and Chromosome Research, and to Genes and Genomes. Secondly, the area of research has also moved from observational to experimental and to analytical during the last decade. Besides, the fact that the citations of more than 77% of references in this book are dated after 2000, clearly indicates that this area of research is not only 'hot' but also 'explosive'.

A reason for the explosive number of publications in this 'hot' area may be traced to the availability of new techniques and tools. Table 6 briefly explains some of these techniques as an introduction to graduate students. The *in situ hybridization* technique, first described in 1969, bloomed into FISH in 1990. Nanda et al. (2000) are perhaps the first to use FISH with BACs as a probe to detect sex chromosomes in *Xiphophorus maculatus*. Immuno-histochemical staining has been widely used in this area of research; for instance, adaptive

Table 6. Development of techniques employed for understanding sex differentiation genes in fishes

Technique	Reference	Principle/Advancement over the earlier one
in situ hybridization (ISH)	Pardue and Gall (1969)	ISH is a type of hybridization that uses a labeled complementary DNA or RNA strand (i.e., probe) to identify or localize a specific DNA or RNA sequence in a section of tissue (*in situ*) or in the entire tissue (whole mount ISH). DNA ISH determines the structure or integrity of chromosomes. RNA ISH quantitates and localizes mRNAs within tissue sections or whole mounts.
fluorescence in situ hybridization (FISH)	Lengauer et al. (1990)	FISH, a cytogenetic technique, detects and localizes the presence or absence of specific DNA sequences on chromosomes. It uses fluorescent probes that bind to only those parts of the chromosome with which they show a high degree of nucleotide sequence homology. It is used to find specific features in DNA for use in genetic counselling, medicine, and species identification. It can also quantitate and localize specific mRNAs within tissue and also analyzes the spatial-temporal patterns of gene expression within cells and tissues.
Immuno staining/Immuno-chemical staining/ Immunohisto-chemistry (IHC)	Coons et al. (1941)	IHC staining of tissue sections or cells is one of the most commonly used immunostaining techniques and uses an antibody-based method to detect a specific protein in a sample. It is extensively used in basic research to understand the distribution/ localization of biomarkers or differentially expressed proteins in different parts of a biological tissue. Visualization of an antibody-antigen interaction is accomplished in a number of ways. Otherwise, the antibody can also be tagged to a fluorochrome such as fluorescein. While the first cases of IHC staining used fluorescent dyes, other non-fluorescent peroxidase and alkaline phosphatase are presently used. These enzymes are capable of catalyzing reactions that provide light microscopically detectable colored products. Alternatively, radioactive elements are used as labels, and the immunoreaction is visualized by autoradiography.
PCR	Mullis and Faloona (1987); Mullis et al. (1986), see also Saiki et al. (1988)	Used to amplify a single or a few copies of a piece of DNA across several orders of magnitude, generating thousands to millions of copies of a particular DNA sequence. All PCR applications employ a heat-stable DNA polymerase, such as Taq polymerase, an enzyme originally isolated from the bacterium *Thermus aquaticus*. This DNA polymerase enzymatically assembles a new DNA strand from single-stranded DNA as a template and DNA oligonucleotides (also called DNA primers), which are required for initiation of DNA synthesis. The vast majority of PCR methods use thermal cycling, i.e., alternately heating and cooling the PCR sample to a defined series of temperature steps.

Table 6. contd....

Table 6. contd....

Real-time polymerase chain reaction (Real time PCR/RT PCR)	Higuchi et al. (1992)	It is also called *quantitative real time polymerase chain reaction* (Q-PCR/qPCR/qRT-PCR) or *kinetic polymerase chain reaction* (KPCR), amplifies and quantifies a targeted DNA molecule, enables both detection and quantification of the amplified DNA as the reaction progresses in *real time.* It quantitatively measures starting amounts of DNA, cDNA, or RNA. Quantitative real-time PCR has a very high degree of precision, e.g., mRNA measurement in single cells (Bengtsson et al., 2008). QRT-PCR methods use fluorescent dyes, such as Sybr Green, EvaGreen or fluorophore-containing DNA probes, such as TaqMan, to measure the amount of amplified product in real time.
Rapid Amplification of cDNA Ends (RACE PCR)	Sambrook and Russell (2001)	This technique is used to obtain the full length sequence of an RNA transcript found within a cell. RACE results in the production of a cDNA copy of the RNA sequence of interest and is produced through reverse transcription, followed by PCR amplification of the cDNA copies. RACE can provide the sequence of an RNA transcript from a small known sequence within the transcript to the 5′ end (5′ RACE-PCR) or 3′ end (3′ RACE-PCR) of the RNA. This technique is also called as *one-sided PCR* or *anchored PCR.*
Microarray	Augenlicht and Kobrin (1982)	This technology evolved from Southern blotting, where fragmented DNA is attached to a substrate and then probed with a known gene or fragment. The first reported use of this approach is the analysis of 378 arrayed bacterial colonies each harboring a gene. Subsequently this technology has been extensively evolved and current versions can analyze more than a million genomic regions in a single experiment. Three major types of arrays are: expression arrays, copy number arrays and SNP arrays. The probes are attached on glass slides and hybridized with the RNA/DNA sample of interest and thus the expression status of the genomic regions spotted on the microarray chips are inferred.
Trangenesis	Palmitter et al. (1982)	Successful production of transgenic mouse by introducing metalothionein-human growth hormone fusion gene (*hGH*) into mouse egg resulting in dramatic increase in growth. The restricted scope of cytoplasmic introduction of transgene is one of hurdles in transgenic fish production. The use of triploid fish eggs avoids the propagation of escaped transgenics (Pandian and Marian, 1994).

immune cells have been detected in the sex changing gonads of the sea bream *Sparus aurata* (Liarte et al., 2007).

The discovery of PCR in 1986 is a land mark event. Since its discovery and patenting in 1987 (Process for Amplifying Nucleic Acids, US 4,683 dated July 28, 1987), the machine has been developed into Real Time PCR in 1992 and RACE PCR in 2001. Yoshizaki et al. (2000), Kwon et al. (2001a,b) and Guo et al. (2008b) are perhaps the first to use PCR, RT PCR and RACE PCR in this area of research, respectively. For instance, Yoshizaki et al. have used a reverse transcription PCR analysis to show that the *Vasa* of rainbow trout is maternally inherited. Since its introduction a little more than 15 years, RT PCR has become the main technical platform for nucleic acid detection in research and development (Kubista, 2008). At present almost all authors working in the area of sex differentiation in fishes seem to employ the RT PCR (e.g., Tse and Ge, 2010; Gunter et al., 2011; Zhang et al., 2011).

Over the past 10 years, DNA microarrays have been used for the analysis of gene expression (Goetz and MacKenzie, 2008). Microarrays have been successfully employed by Santos et al. (2007) for the first time in this area of research. Santos et al. have provided the first description and analysis of the reproductive gonadal transcriptome of an actively breeding zebrafish. Using affymetrix microarrays to gene expression patterns in male and female somatic and gonadal tissues, Small et al. (2009) have reported the differences in transcript abundance of 5,899 genes between male and female zebrafish. All sex-biases in gene expression are due to the difference between testes and ovaries.

There are many publications that discuss the potential benefits and possible environmental hazards of transgenic fishes (e.g., Zbikowska, 2003; Murray and Maga, 2010). However transgenics are also generated for the purpose of understanding the sex differentiation process. Krovel and Olsen (2002, 2004) have found that the *Vas : : egfp* transgene is strongly expressed in gonads of the female but only faintly in that of male zebrafish.

2

Differentiation in Gonochores

Sex is a luxury and costs time and energy but ensures recombination to generate genetic diversity (Carvalho, 2003). As benefits accruing from genetic recombination outweigh the costs of time and energy, sex is successful and has evolved as early as 1.6–2.0 billion years ago (Butlin, 2002) and has been successfully manifested in a wide range of microbes, plants and animals. Of the 30, 000 fish species, more than 98% are bisexuals and the others are non-bisexuals, i.e., hermaphrodites or unisexuals. Even among the hermaphrodites, hardly two species are self-fertilizing; all other hermaphrodites function as male or female at a given point of time/age (Pandian, 2011). Importantly, the sex changing sequentials provide a distinct opportunity to study the processes of sex differentiation, de-differentiation and re-differentiation over a period of time.

2.1 Sex chromosomes

Sexuality in fishes ranges from gonochorism to unisexualism and self-fertilizing hermaphroditism to sequential and serial hermaphroditism, suggesting the presence and operation of diverse sex determination mechanisms. Cytological mechanisms of hereditary transmission by chromosomes as carriers of genes have received much attention. Yet our present knowledge is based on cytogenetic characterization of only about 1,700 species, of which 188 species belonging to 86 families, i.e., 6–11% of fishes alone are reported to have cytogenetically differentiated heteromorphic sex chromosomes (Devlin and Nagahama, 2002). The use of sensitive techniques like the anti-body staining of Human MutL Homologue 1 (MLH1) protein to mark the sites of putative chiasmata, the physical basis for the pattern of recombination, opens a new avenue to recognize the presence of sex chromosomes; for instance, the ratio of MLH1 loci per chromosome in males versus females is 1 : 1.55 indicating the presence of sex chromosome in zebrafish (Kochakpour and Moens, 2008), in which no heteromorphic sex chromosomes could earlier be recognized (e.g., Wallace and Wallace, 2003). Of the 13,000 marine fish species, fewer than 2% have so far received cytogenetic attention (Brum, 1996). Of the 400 and odd species displaying one or another form of hermaphroditism, sex chromosomes have been hardly recognized in a half dozen species. The Neotropical fishes are of special interest, as about 8,000 freshwater fishes, representing more than 25% of all fish species are considered to be present there (Schaefer, 1998). Of the 4,475 species thus far recognized, the presence of sex chromosomes has been claimed only for 5.9% species (Oliveira et al., 2009). Using more sensitive chromosome painting and other techniques, the presence of sex chromosomes in another half a dozen Neotropical *Characidium* has been recently reported (Machado et al., 2011). Nevertheless, we seem to have just touched the tip of the iceberg and our basic understanding of cytogenetics is limited to a few thousand species and much is awaited especially from the coral-reef inhabiting hermaphroditic and Neotropical fishes.

A heterogametic individual produces genetically different gametes (X, Y or Z, W), one of which may determine the sex of its progeny generated by its fusion with the gamete of the opposite sex. On the basis of reliable evidences from heteromorphic sex chromosomes, 264 fish species belonging to about 160 families are recognized as heterogametic; of them, 67% are recognized as male heterogametic (Pandian, 2011). Male and female heterogametics also exist within a genus (e.g., *Gambusia*, *Xiphophorus*) and within a species (*Gambusia affinis*) too. More intriguingly, 9% fish species are known to have multiple sex chromosome-systems. There are frequent reports on the existence of multiple sex chromosomes, especially from the Neotropic freshwater fishes and endemic species from the Antarctic (Pandian, 2011). The presence of male and female heterogametic sex determination systems as well as simple and multiple sex chromosomes clearly indicates the absence of a common, universal sex chromosome system in fishes, as mammals have it. It is not known whether these diversities in gamety and simple/multiple sex chromosome systems are reflected in sex differentiation or not. For instance, high androgen ratios in male heterogametic vertebrates may be the consequence of the Y-linked gene, whose product is an inhibitor of *cyp19*, whereas the W chromosome may carry a gene inducer of *cyp19* in female heterogamatics (Bogart, 1987; see also Piferrer and Guiguen, 2008)

The family Cichlidae, one of the most species-rich families of vertebrates, is composed of 3,000 species. More than 1,500 species inhabiting the Great Lakes of eastern Africa are endemic. Because of their impressive phenotypic diversity, explosive speciation rate, and unique endemism, the African cichlids have attracted the attention of researchers (Gunter et al., 2011). More than 1,000 species of cichlids representing about 10% of the world's freshwater species, have arisen within the past million years in Lakes Malawi and Victoria in eastern Africa (Allender et al., 2003). The mouthbrooding tilapias also share a common and more recent ancestry of less than 5 million years (My). Hence the differentiation process of sex chromosome is of interest. Gnaani et al. (2008) identified DNA markers linked to sex determining genes in six closely related species of tilapias. In the female heterogametic *Oreochromis karongae* and *Tilapia mariae*, the sex determining locus is on linkage group LG3 but it is in LG1 in male heterogametic *O. niloticus* and *T. zilli* (Table 7). Comparison of genetic and physical maps suggests relatively stronger sex specific suppression of recombination due to accumulation of repetitive elements in the sex determining region in female heterogametic *O. aureus* than in male heterogametic *O. mossambicus*; notably the W and Z homologues of LG3 in *O. aureus* remain cytologically indistinguishable, suggesting one or more inversions in the W chromosome.

Table 7. Known linkage groups and sex chromosome systems in selected fish

Species	Sex chromosome system	Linkage group (LG)	Reference
Gasterosteus aculeatus	XX/XY	19	Ross and Peichel (2008)
Pungitius pungitius	XX/XY	12	Shapiro et al. (2009)
Oreochromis niloticus	XX/XY	1	Gnaani et al. (2008)
Tilapia zilli	XX/XY	1	Gnaani et al. (2008)
O. karongae	ZZ/ZW	3	Gnaani et al. (2008)
T. mariae	ZZ/ZW	3	Gnaani et al. (2008)
O. mossambicus	XX/XY	3	Gnaani et al. (2008)
O. aureus	ZZ/ZW	3	Gnaani et al. (2008)
Takifugu rubripes	XX/XY	19	Kikuchi et al. (2007)
Oryzias latipes	XX/XY	1	Tanaka et al. (2007)
O. curvinotus	XX/XY	1	Tanaka et al. (2007)
O. luzonensis	XX/XY	12	Tanaka et al. (2007)
O. mekongnensis	XX/XY	2	Tanaka et al. (2007)
O. minutillus	XX/XY	8	Tanaka et al. (2007)
O. dancena	XX/XY	10	Tanaka et al. (2007)
O. hubbsi	ZZ/ZW	5	Takehana et al. (2007)
O. javanicus	ZZ/ZW	16	Takehana et al. (2007)
Poecilia reticulata	XX/XY	12	Watanabe et al. (2005)
Xiphophorus maculatus	XX/XY	24	Bohne et al. (2008)
Salmo salar	XX/XY	2	Artieri et al. (2006)

Constructing a medium-density genetic linkage map of microsatellite markers of *Astatotilapia burtoni*, Sanetra et al. (2009) showed the presence of 25 linkage groups spanning 1249.3 cM of the genome size (~ 950 Mb) with an average marker spacing of 6.12 cM. The seven *Hox* clusters, *ParaHoxC1* and two paralogues of *Pdgfrβ* have also been mapped to different linkage groups; their findings support the hypothesis of teleost-specific genome duplication, with the formation of an astonishingly large number of species in a record-breaking time in cichlids.

Amazingly, among the Malawi cichlids investigated by Ser et al. (2009), seven species have XX/XY system and another three possess ZZ/ZW system (Table 8). Evidences from several species strongly suggest the presence of additional sex determining loci and cryptic epistatic interactions. For instance, the known two sex determining XX/XY and ZZ/ZW systems co-exist and segregate within single families of *Metriaclima pyrsonotus*. When both dominant sex determinant loci W and Y are present in a single individual (Table 8), the ZZ/ZW system dominates over the XX/XY system.

Table 8. Known linkage groups and sex chromosome systems in selected Malawi cichlids (compiled from Ser et al., 2009)

Species	Sex chromosome system	Linkage group (LG)
Aulonocara baenschi	XX/XY	7
Pseudotropheus polit	XX/XY	7
Metriaclima barlowi	XX/XY	7
M. mbenjii	XX/XY	7
M. phaeos	XX/XY	7
M. zebra	XX/XY	7
M. pyrosonotus Family Mlpy A002, A017 Family Rmpy 018, 021	XX/XY ZW/XY	7 5
M. calloinos	ZW/XY	5
M. fainzilberi	ZW/XY	5
Labeotropheus trewavasae	ZW/XY	5

Uniquely, the platyfish *Xiphophorus maculatus* has three different, genetically defined sex chromosomes X, Y and W (Kallman, 1984). Possible female genotypes are XX, XW and YW, whereas males harbor the XY or YY constellations. Unlike in mammals, YY male and ZZ female fishes are viable and fertile (Pandian, 2011). WW females have not been recorded from natural populations but can be generated in the laboratory and are viable as well (Schultheis et al., 2009). Sex determining mechanisms among the many species of *Xiphophorus* are quite diverse and have been well established in 12 species (Walter et al., 2006). *X. maculatus* Jp 163A is a female homogametic and a cross between it and *X. helleri* Sarabia male produces homogametic YY male progenies. Hence *X. maculatus* Jp 163A strain is regarded as an ideal model to understand the sex determining/differentiation processes. Employing the 454–FLX massively parallel DNA sequencing platform, Zhang et al. (2011) have obtained 742,771 and 721,543 reads from whole adult female and male, respectively. Among the identified genes 54 are differentially expressing sex genes. Further, the Poisson-based enriched testing has identified 2,250 male-enriched and 2,304 female-enriched contigs with various degrees of difference. Among these contigs, 22 and 23 are female-and-male predominant transcripts, respectively. Within these contigs, eight female predominant and nine male predominant transcripts are recognized as differentially expressing sex genes.

Even among these 6–11% fish species, in which the existence of sex chromosomes has been recognized, many claims are adduced from genetic rather than cytological evidence. Emphasizing the need for molecular cytogenetic approach, Ross and Peichel (2008) showed the presence of sex chromosomes in the stickleback *Gasterosteus aculeatus*, in which the sex chromosomes are morphologically homomorphic. A BAC (Bacterial Artificial Chromosome) containing *cyp19b*, located at 16.7 Mbp in the X chromosome assembly, hybridizes only to the X but not to the Y nor to any other location in the genome, suggesting that a part of the Y chromosome has been deleted (Fig. 5). The conclusion is supported by the presence of *cyp19b* on the X but not on Y, even when detected with other probes around *cyp19b*. This large (6 Mbp) deletion on the Y is equivalent to 30% of the sequence content of X chromosome. However, X and Y chromosomes are morphologically similar in size at metaphase. This is due to a pericentric inversion resulting in difference in the centromere positions of X and Y chromosomes. Comparing a few hundred thousand base pairs of sequence from the non-recombination region of the Y chromosome and homologous region of the X chromosome, Peichel et al. (2004) have shown the accumulation of many sequence characteristics in the sex chromosome, including an elevated transposable element content and a small intra-chromosomal duplication and inversions (*Wtla*). Briefly, the sex determining region of the Y or Z chromosome is a 'hot spot' for deletion, inversion, duplication, amplication and transposition and other kinds of rearrangements (see Pandian, 2011). For instance, Y choromosome-autosome fusions have occurred in at least 25 fish species, including twice independently in stickebacks (Ross et al., 2009). Hence there is a need for molecular cytogenetic approach to recognize sex chromosomes (see also Volff et al., 2007).

A further complication in sex differentiation is caused by autosomal genes inducing departures from expected sex ratio. Examples for such unexpected departures in sex ratio are plentiful in relevant literature (e.g., Herpin et al., 2007). In the context of gynogenesis and androgenesis, some of

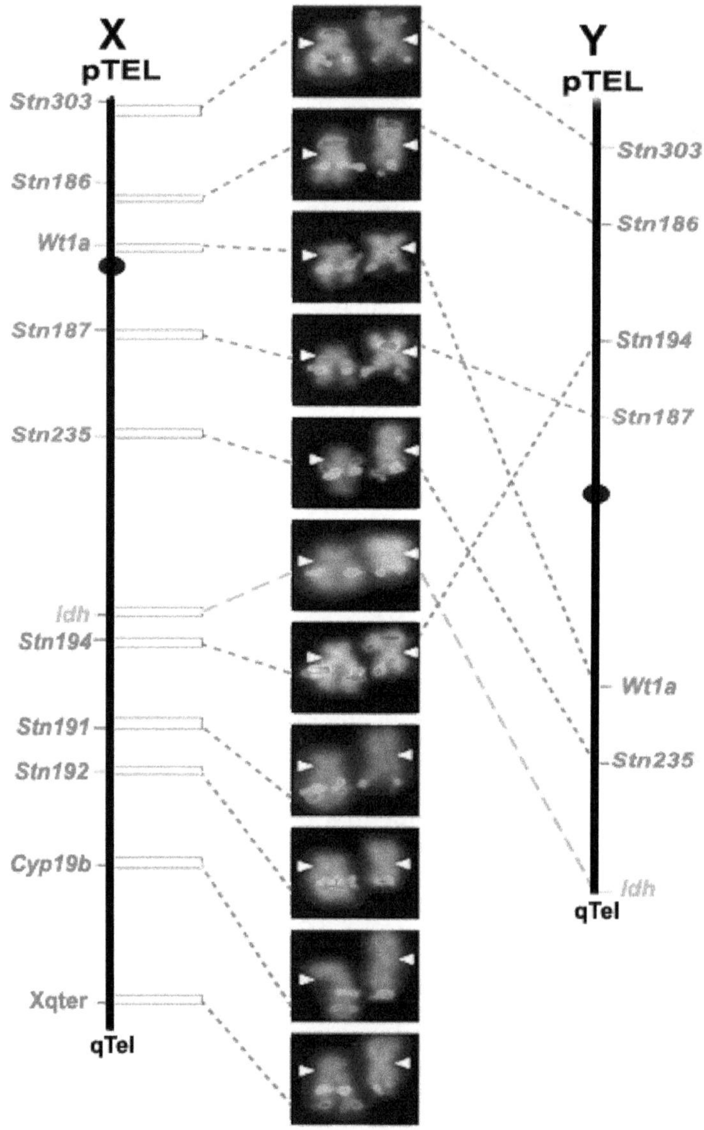

Fig. 5. FISH-based cytogenetic maps of X (left) and Y (right) chromosomes of *Gasterosteus aculeatus*. Each FISH section (center) contains the X and Y from a single metaphase spread, to which *Idh* and an additional probe are hybridized. The position of the centromere is identified with a white arrowhead. Dashed lines joining the X map to the FISH section indicate the BHC probe used in each section and those to the Y cytogenic map indicate the physical interval to which each FISH probe hybridizes. The black circles represent the positions of the centromeres. Note the deletion of *cyp19b* in the Y (from Ross and Peichel. 2008, with permission by Genetic Society of America)

Color image of this figure appears in the color plate section at the end of the book.

these unexpected departures in sex ratio are briefly summarized by Pandian (2011). PCR analysis has shown that the unexpected female progenies sired by YY androgenics are indeed genetic males but function as female phenotypes (e.g., *Puntius conchonius*, Kirankumar et al., 2003; *Hemigrammus caudovittatus*, David and Pandian, 2006; Fig. 6). In *O. latipes*, a cross between two strains namely ($Y^{Awr}Y^{Awr}$) ♀ x Ol-vas (XY^{Hdr}) ♂ is also reported to produce Ol-vas ($Y^{Awr}Y^{Hdr}$) ♂ and Ol-vas (XY^{Awr}) ♀. But in the absence of *Dmrt1bY*^{Hdr} expression, the cross produces genotypic and phenotypic females. In the presence of Ol-vas genotype, a phenotypic hermaphrodite and in the absence of it, phenotypic males are also generated, as shown by PCR analysis (Herpin et al., 2007). Hence PCR analysis to identify genetic sex of

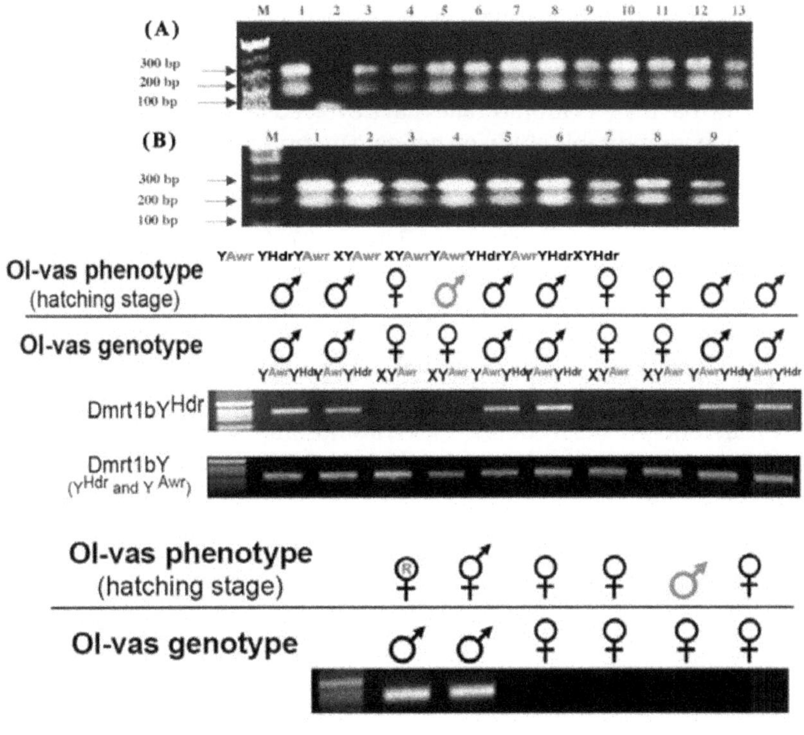

Fig. 6. Upper panel: PCR products of the genomic DNA of *Hemigrammus caudovittatus* as amplified by Dmrt1 primer. A row: Lane 1 = male, Lane 2 = female, Lanes 3–13 androgenic males generated using fresh or cadaveric sperm. B row: Lanes 1–7 unexpected female progenies and Lanes 8 and 9 male progenies (from David, 2004). Lower panel: *Oryzias latipes* progenies sired by an Ol-vas XY^{Hdr} ♂ crossed with $Y^{Awr}Y^{Awr}$ female. Note the absence of *Dmrt1bY*^{Hdr} expression on genotypic and phenotypic females and also the absence of expression of Ol-vas in genotypic and phenotypic females but its presence in males including a phenotypic hermaphrodite (from Herpin et al., 2007)

Color image of this figure appears in the color plate section at the end of the book.

all such ambiguous fishes has become necessary. These findings implicate that the function/expression of one or more genes in genetic cascade of sex determination system is altered resulting in the reversal of the genetic sex to the opposite phenotypic sex.

Possession of a genotype of one sex and phenotypic expression of the opposite sex by an individual clearly indicates the need for sex specific molecular markers in fishes. Many molecular markers are described to identify genetic sex of a few fishes. A series of sex-linked molecular markers have also been described, for instance, in many salmonids and poeciliids (Table 9). The Y chromosome linkage maps of four salmonids *Salvelinus alpinus, Salmo salar, S. trutta* and *Oncorhynchus mykiss* reveal the telometric placement of the sex determining region involving a smaller fraction of chromosome in the first three species but an intercalary position for *O. mykiss*. In each species, the Y specific region is very short and the major male determining gene is located on different linkage groups in these salmonids, indicating that their Y chromosomes have evolved independently (Woram et al., 2003). In a creditable study on genome mapping of the sex determining region in sex chromosomes with several different molecular makers, Artieri et al. (2006) have identified the location of the sex determining gene SEX on the metacentric arm of chromosome 2 in *S. salar.* After genome duplication events 360 Mya (see Pandian, 2011), many salmonids have dealt with sex determination by deleting one copy of the sex determining locus; for instance, *S. salar* deleted one copy SEX in females, growth hormone pseudogene *GH-ψ* in *O. masou* and *O. tshawytscha* (Davidson et al., 2009; see Table 9). In another remarkable investigation in *X. maculatus,* in which three different genetically defined sex chromosomes X, Y and W are present, Schultheis et al. (2006) and Bohne et al. (2008, 2009) have located the master sex determining gene in the Y chromosome between the melanoma-inducing oncogene *Xmrk* and its proto-oncogene counterpart *egfrb*. The known linkage groups and sex chromosome system (Tables 7, 8, 9) in selected fish (for more details see Graves and Peichel, 2010) provide many examples for "the repeated emergence of new sex chromosomes from autosomes and thus bringing the sex differentiation cascade under new master genes". In view of its importance, several projects are now undertaken to clone the master sex determining genes identified so far in stickleback, fugu, tilapia, salmonids, guppy and platyfish (see Schultheis et al., 2009).

2.2 Sex determining genes

In vertebrates only two sex determining genes have thus far been identified; the mammalian *SRY* gene and *DMY/Dmrt1bY* gene in medaka fishes *O. latipes* and *O. curvinotus* (Matsuda et al., 2002; Nanda et al., 2002). *SRY* and *DMY* are not homologous genes. *SRY* encodes a transcription factor

Table 9. Representative examples of molecular markers to identify sex, sex chromosome and sex determining locus in selected fishes (from Pandian, 2011, updated)

Species and Reference	Remarks
Xiphophorus maculatus Coughlan et al. (1999)	PCR based genomic DNA analysis identified the presence of 323bp fragment in XX and YZ females and 323 bp plus high molecular fragment in XY and YY males
Puntius conchonius Kirankumar et al. (2003)	*SRY* specific primer identified male specific 588, 333 and 200 bp fragments in males but only 200 bp fragment in females
Hemigrammus caudovittatus David and Pandian (2006)	DMRT-1 specific primer identified 237 and 300 bp fragments in males but a 200 bp fragment in females
Takifugu rubripes Cui et al. (2006)	Sex linked anonymous marker sequence suggested female heterogamety
Clarias gariepinus Kovacs et al. (2000)	Sex linked RAPD marker *CgaY1* (2.6 kb) present in males and produced faint male specific band at low stringency in *Heterobranchus longifilis*. *CgaY2* (458 bp) produced similar hybridization pattern in both sexes of *C. gariepinus*, *C. macrocephalus* and *H. longifilis*
Oncorhynchus spp Du et al. (1993), Nakayama et al. (1998), Devlin et al. (2001)	Sex linked *GH* pseudogene identified in *O. keta*, *O. gorbuscha* and *O. kisutch*, but it is absent in *O. nerka*, *O. mykiss* and *O. clarki*
O. masou Zhang et al. (2001)	PCR based identification of growth hormone pseudogene *GHψ* in 93–98% of males and in a few females, whose *GH-ψ* fragment was identical to that of males and shared 95% homology with that in male *O. tshawytscha*
O. tshawytscha Nagler et al. (2004)	Real time quantitative PCR analysis detected the presence of a single copy of growth hormone pseudogene *GH*p in males but its absence in females
O. tschawytscha Brunelli and Thorgaard (2004)	AFLP analysis of purified DNA of YY gynogenic ♂ and ♀ identified OtY2 that was amplified in ♂ but not in ♀. OtY2 androgenic male specific amplification were present in coho, chum and sockeye salmon but not in pink salmon and rainbow trout
Seriola quinqueradiata Fuji et al. (2010)	Using 71 microsatellite markers, selected from genetic linkage map of female (ZW), detected the presence of sex determining locus between *sequ21* and *sequ17* in LG12
Salvelinus namaycush Stein et al. (2002)	Detection of sex linked microsatellite locus near the telomere of Y chromosome
X. helleri Kazianis et al. (2004), Woolcock et al. (2006)	Identification of sex chromosomes by several molecular markers and mapping of sex determining region by linkage group 24
Poecilia reticulata Kirpichnikov (1981), Khoo et al. (1999a, b), Lindholm et al. (2004), Nanda et al. (1992), Tripathy et al. (2009)	Traits like Black caudal peduncle (Bcp) and Red tail (*Rdt*) markers linked to the dominant T chromosome. Traits like *Macropodus*, *Aramatus* and *Pauper* pigmentation linked to YY chromosomes. Using $(GATA)_4$ and $(GACA)_4$ probes, sex chromosomes were identified. Genetic linkage map of the genome using 790 single nucleotide polymorphism markers
Verasper variegattus Ma et al. (2010)	V'eval F 533 and species specific V'eval 218 markers present in females only

containing a high mobility group (HMG) box domain but the *DMY* is a transcription factor similar to *the double-sex and male-3 related transcription factor* (*Dmrt1*). Another candidate gene, the anti-Mullerian hormone *amh* has been mapped but within an autosomal sex determining Quantitative Trait Loci (QTL, Shirak et al., 2006). Using a plethora of traditional molecular approaches such as AFLPs, RAPDs and subtractive hybridization, the search for sturgeon sex determining gene(s) has remained unsuccessful (e.g., McCornick et al., 2008), as the sturgeon genome is polyploid and most loci are inherited tetrasomically or octosomically (Blacklidge and Bidwell, 1993). However, Hale et al. (2010) claim the discovery of candidate sex-determining genes *Dmrt1* and *Tra-1 Dmrt1* in *Acipenser fulvescens.* Incidentally, *Dmrt1* is the master sex determining gene in birds and medaka, whereas *Tra-1* helps direct sexual differentiation in nematodes. Hale et al. have employed more advanced methods like 454 pyrosequencing, expressed sequential tags, ontologies and so forth, used to characterize non-model species bearing homomorphic chromosomes. They have also attempted to adduce evidence with *Trichomonas* BLAST hits limited to the female gonad. However, they have not located the claimed sex determinant genes to a specific sex chromosome or autosome. The claims by Shirak et al. (2006) and Hale et al. (2010) need to be confirmed in other cichlid and acipenserid species.

Employing a novel procedure to study genome-wide linkage, Bradley et al. (2011) have also claimed the discovery of sex determination genes in zebrafish, in which the identification of sex chromosomes and sex linked genes has remained elusive (e.g., Wallace and Wallace, 2003). Genotyping, genome-wide linkage analysis and recombination estimates of an accurately developed SNP (Single Nucleotide Polymorphism) genetic map of zebrafish have shown that sex determination is indeed a complex trait in zebrafish and does not employ sex chromosomes in the conventional homogametic/heterogametic mode. Bradley et al. have estimated the LOD scores for linkage to sex determination, which have identified two loci, that exceed genome-wide significance for linkage to sex determination. One locus resides on chromosome 5 at 53.5 cM with an LOD score of 7.9 and the other on chromosome 16 at 32.0 cM with an LOD score of 9.3. Further analysis of mutants has suggested that the *Dmrt1* and *cyp21a2* are the candidate genes harbored in chromosomes 5 and 16, respectively. Mutagenetic data, that defined the highly conserved 3′ UTR protein binding motif shown to be responsible for transcription stability and translation efficiency, predict that the female associated allele 5′ CUGCUACAGAU-3′ would yield to lower *Dmrt1* expression relative to male allele 5′ CUGCUGCAGAU-3′ in the developing gonadal primordium. *cyp21a2* encodes 21-hydroxylase biosynthesizing corticosteroids. Reduced 21-hydroxylase activity results in shunt of precursors toward steroid biosynthesis. Corticosteroid as well as higher temperature can induce female to male reversion (however, see also Frisch et al., 2007).

DMY and its paralogue *Dmrt1*bY gene, found only in *O. latipes* and *O. curvinotus*, is a duplicated copy of the autosomal region, which spans to 258Kb and contains only one functional *DMY* gene. This Y specific region has originated from a transposition of the duplicated autosomal region (Matsuda et al., 2002; Nanda et al., 2002). A single amino acid change from *Thr* in the three sequences to *Ser* in the four *Dmrt1* sequences has been largely responsible for the establishment of *DMY/Dmrt1bY* as male sex determination gene (Zhang, 2004). This Y chromosome specific *DM* domain gene is required for male development.

The strain Hd-rR^{HNI} of *O. latipes*, a hybrid strain generated using a strain carrying sex determining region^{HNI} and Hd-rR strain carrying the R dominant red body color gene on the Y chromosome and its allele the r for the recessive white body color (Shinomiya et al., 1999). Matsuda et al. (2007) sub-cloned the DMY^{HNI} genomic region of *O. latipes* to the *p*CC1 vector from BAC. The vector containing DMY^{HNI} genomic fragment was injected into one-cell stage embryos of medaka d-rR strain. Expectedly, 57 orange-red adult male (XY) and 58 white (XX) adult females were obtained in the F_1 generation. Genomic PCR-RFLP analysis revealed that both transgenic and non-transgenic individuals, which carried the sex determining region with DMY^{HNI} differentiated into males. However, due to overexpression of *DMY*, some of the transgenics, that carried the sex determing region namely DMY^{HNI} became either fertile or sterile females (Fig. 7). Of the 58 white XX fish, 13 (22%) displayed secondary sex characteristics of male. Further overexpression of *DMY* cDNA also caused XX sex reversal to male, as indicated by the presence of acinus in some and delayed developing ovaries with reduced number germ cells or with incomplete follicles in others. These findings demonstrate that *DMY* is sufficient for male development in medaka and suggest that the functional difference between the X and Y chromosome is a single gene. Incidentally, YY female medaka with mutant-type of *DMY* genes on their Y chromosomes are completely fertile, indicating that the X and Y chromosomes are functionally the same except for the male determining function (Otake et al., 2008).

2.3 Sex differentiation genes

In an attempt to survey the genes that are involved in sex differentiation in fish, relevant information is compiled from the reviews of von Holfsten and Olsson (2005), Pala et al. (2009), Sandra and Norma (2009) and Siegfried (2010).

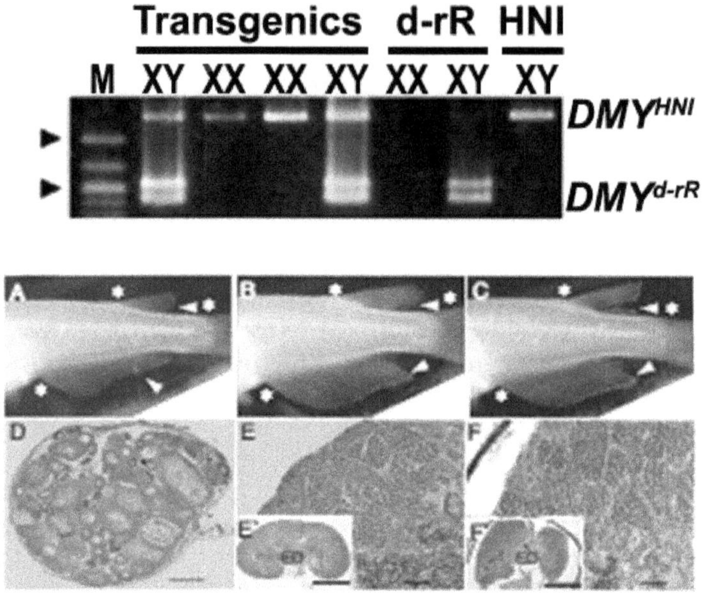

Fig. 7. *Oryzias latipes*: Upper panel: M = DNA marker. Transgenic XX individuals have only the Dmy^{HNI} allele, whereas XY individuals have both Dmy^{HNI} and $Dmy^{d\text{-}rR}$ alleles. Lower panel: First row: The secondary sexual characteristics and body coloration of the *Dmy* transgenics. Note the larger size and shape of dorsal and anal fins (white arrow heads) in males. Second row: Histologic sections of gonads. A and D = XX female; B and E-E^1 = XX medaka with testis and C and F-F^1 XY male. ED = efferent duct. Scale D, E^1 and F^1 = 200 µm E and F = 50 µm (from Matsuda et al., 2007. *Dmy* gene induces male development in genetically female (XX) medaka fish. Proc Natl Acad Sci, USA, 104 : 3865 – 3870. Copyright 2007 National Academy of Sciences, USA)

(i) *sox9*: The *sox* family of genes encodes transcription factors containing the *Sry* related HMG box, a DNA binding motif of approximately 79 amino acids. At least 20 *sox* genes have been identified and are subdivided into 10 groups A to J based on their sequences and are involved in a wide range of developmental process including sex determination and neurogenesis (Bowles et al., 2000). *sox9* is classified into group E and has a transactivation domain in its C terminus (Sudback et al., 1996). The *sox9* identified in medaka encodes 487 amino acids and shows 70% amino acid identity with known vertebrate sox9 proteins. In zebrafish (Chiang et al., 2001) and medaka (Yokoi et al., 2002), *sox9a* is expressed in the testes but *sox9b* in the ovary. In *O. mykiss*, *sox9a1* shows sexually dimorphic expression (Vizzanio et al., 2007). The full length of *sox9b* of guppy *Poecilia reticulata* coding sequence is 1,443 bp and encodes for a protein of 480 amino acids. The *sox9b* of the guppy has 88% and 91% similarity to those of *G. aculeatus* and

O. latipes, respectively. It shows no difference in its expression pattern in the male and female guppy but very interestingly a clear expression in muscle of adult males but not in that of females (Shen et al., 2007). Table 10 summarizes the timing and levels of expression of *sox* gene complex in selected fish.

Table 10. Summary of *sox* gene complex in downstream sex differentiation in fishes (from Pandian, 2011, updated)

Species, Reference	Description
Acipenser sturio Hett et al. (2005)	*sox* 9 consists of 3 exons and 2 introns with completely conserved exon-intron boundaries and shows high levels of homology to *sox9* sequences of other vertebrates, especially in the N-terminus region containing the HMG box. It reveals no male or female specific difference
A.schrenckii Chen et al. (2006)	*sox* genes express differently in males and females. They code for transcription factors related to mammalian *Sry* testis-determining gene
Cyprinus carpio Du et al. (2007)	*sox9a* and *sox9b* differ in having an intron of different lengths of 704 and 616 bp, respectively in the conserved HMG box region that codes for identical amino acid sequence. C*csoxb* overexpresses in adult brain and testis but weakly in ovary
Allotetraploid Liu et al. (2007)	*sox9a* is highly expressed in testis but weakly in brain and heart, and is not expressed in adult female
3n crucian carp Guo et al. (2008b)	Isolated and characterized cDNA of *sox9a*, encodes a protein of 457 amino acids with an HMG box. *sox9a* is expressed in testis, as in zebrafish. But the testicular expression of *sox9a* gene 3n seldom contributes to the formation of functional spermatozoa, although plays an important role in development of testicular tubules. Thus the contribution of *sox9* genes differs in different species
Danio rerio von Hofsten et al. (2005a,b)	Two *sox9* genes, *sox9a* and *sox9b* have been identified. Both contain HMG box and are able to bind the AACAAAG recognition site. *sox9a* is expressed in testis and *sox9b* in ovary. *FTZ-F1a* (*fushi tarazu* factor-1a) genes are involved in regulation of inter-renal development and thereby steroid biosynthesis. sox9a alone does not determine sex but a combination of *sox9a*, *FF1a, b, c* and c, MMH GATA and *Dmrt* seems to perform it
Clarias gariepinus Raghuveer et al. (2011)	*Sox9a* is exclusively expressed in male gonads

From a PCR survey of *sox* genes in cyprinid tetraploid tor *Tor douronensis*, Guo et al. (2008b) reported the presence of 13 *sox* genes belonging to *soxB*, *soxC* and *soxE* groups. Sequence analysis revealed that most of the nucleotide substitutions between duplicated copies of *sox* genes were caused by the tetraploidization event or their orthologues of other species remained silent substitution. Unexpected variations of the intron length and similarities of the two duplicated copies of *sox9a* and *sox9b* suggest that *T. douronensis* may be allotetraploid.

(ii) *Dmrt*: The minireview of Herpin and Schartl (2011) summarizes information thus far gathered on the early gonadal differentiation in a dozen gonochoric fish. Of five genes, gonad specific expression has been found for *Dmrt1* in many species, *Dmrt2* in *O. latipes* (Winkler et al., 2004), *Dmrt3* in *D. rerio* (Li et al., 2008), *Dmrt4* in *O. aureus* (Cao et al., 2010), *O. latipes* (Winkler et al., 2004), *Paralichthys olivaceus* (Wen et al., 2009) and *Dmrt 5* in *D. rerio* (Guo et al., 2004). Interestingly, *Dmrt4* and *Dmrt1* are predominantly expressed in the ovary and testis of *O. aureus*, respectively. They have been sub-cloned into the vector pMAL-c2x and introduced into the *Escherichia coli* TB1 cell for efficient fusion. After purification and cleavage, Dmrt4 and Dmrt1 proteins have been used to immunize adult rabbits. Western blot analysis has shown that these *Dmrt4* and *Dmrt1* have become highly specific and are expressed only in the ovary and testis, respectively (Cao et al., 2007). Male restricted expression has been shown in *O. latipes, X. maculatus, P. olivaceus, O. niloticus, Gobiocypris rarus* and *Clarias gariepinus* (Table 11). A strong male-biased expression has been found in *O. mykiss, D. rerio, Takifugu rubripes, Gadus morhua, Acipenser fulvescens, Scaphirhynchus platorhynchus* and *Silurus meridionalis*. When detected in the ovary, the expression of *Dmrt1* is specific to germ cells alone in *G. morhua* and *D. rerio*. Two distinct DM domain DNAs *tDMRT1* from the testis and *tDMO* from the ovary are isolated in *O. niloticus*. A male specific motif is absent in *tDMO*. The alternatively spliced male and female types of *double sex tDmrt-1* and *tDMO* cDNAs are encoded by two different genes. The mutually exclusive nature of *tDmrt-1* and *tDMO* expression in the testis and ovary indicates that they both play an important role in gonad development and function. *In vitro* (promoter analysis) studies have revealed that *Dmrt1* inhibits transcriptional activity of *cyp19a1a* and *Double sex and Mab-3* domain is essential to repress basal as well as *Ad4BP/SF-1*-activated *cyp19a1a* in HEK 293 cells. Immunohistochemical analysis has shown that the transgenic overexpression of *Dmrt1* in XX tilapia down- regulated *cyp19a1a* expression, decreased aromatase expression reduced serum estradiol 17β levels, retardation of the formation of ovarian cavity induced varying degrees of follicular degeneration and even a partial or complete sex reversal (Wang et al., 2010). In *S. meridionalis* too, *Dmrt1a* expresses in the testis but *Dmrt1b* in the ovary (Liu Z et al., 2007). In *Clarias gariepinus,* Raghuveer and Senthilkumaran (2009) have found two alternative spliced forms of *Dmrt1*. They code for proteins with 287 (*Dmrt*), 253 (*Dmrt1a*) and 233 (*Dmrt1b*) amino acid residues and are localized in spermatogonia and spermatocytes. The potential role of *Dmrt1* in testicular differentiation, as evidenced

Table 11. Summary of *Dmrt 1* gene complex in downstream sex differentiation of fishes. Ni-information not provided

Species and Reference	Expression and levels	Expression and localization
Oryzias latipes Kobayashi et al. (2004)	Testis	Pre-Sertoli and, Sertoli cells, Spermatogonial supporting cells, epithelial cells of efferent duct
Xiphophorus maculatus Veith et al. (2006)	Testis	Spermatogonia, Sertoli cells
Paralichthys olivaceus Jo et al. (2007)	Testis	Ni
Oreochromis niloticus Kobayashi et al. (2008)	Testis, Sex reversed testis	Sertoli cells, Epithelial cells of efferent duct
Gobiocypris rarus Zhang et al. (2008)	Testis	Ni
Clarias gariepinus Raghuveer and Senthilkumaran (2009) Raghuveer et al. (2011)	Testis Ovo-testis Exclusively in testis	Spermatogonia, Spermatocytes
Oncorhynchus mykiss Marchand et al. (2000)	Testis, ovary, High in testis	Differentiating testis
Danio rerio Guo et al. (2005)	Testis, ovary, High in testis	Spermatogonia, Spermatocytes, Spermatids, Developing oocytes
Takifugu rubripes Yamaguchi et al. (2006)	Testis, ovary, High in testis	Sertoli cells
Scaphirhynchus platorhynchus Amberg et al. (2009)	Testis, ovary, Higher in testis	Ni
Acipenser fulvescens Hale et al. (2010)	Testis, ovary, High in testis	Ni
Gadus morhua Johnson et al. (2010)	Testis, ovary, High in testis	Germ cells, during spermatogenesis
Silurus meridionalis Liu et al. (2010)	Testis, ovary, High in testis	Ni
Odontesthes bonariensis Fernandino et al. (2008)	Primordial gonads, Testicular differentiation	Ni

from its stage-dependent elevated expression in testes, has been demonstrated perhaps for the first time in a seasonal spawner. Incidentally, seasonal reproductive cyclicity including regeneration of gonadal tissues is mediated by serum hormone levels, brain hormone content and brain enzyme activity, resulting in significant enrichment of neuroendocrine-related genes through the axis of spring-summer (monsoon in Asian countries)–winter. The storage of PGCs mostly singly or in small groups attached to the stroma of the ovary of polycyclic fishes is reported (e.g., viviparous clinid *Fuconimus mus*, Moser, 2007). Analyzing the multiple brain microarray datasets of female *Carassius auratus*, Zhang et al. (2009) have characterized global gene expression patterns as a function of the spawning cycle. A core set of 873 genes in hypothalamus are differentially expressed during the periods of pre-spawning (May), post-spawning (i.e., gonadal regression season in October) and in early gonadal redevelopment (December). Changes in expression pattern of these genes (including *isotocin, ependymin II,* $GABA_A$ *γ2 receptor, calmodulin* and *aromatase b*) are shared by telencephalon. These genes are regulated by photoperiod. The hormones and their genes functionally involved with G-protein coupled receptor, signaling the shift from gonadal regression to the pre-spawning stage.

(iii) Anti-Mullerian hormone (*amh*), also known as Mullerian duct inhibiting substance (*MIS*), is a glycoprotein hormone gene of the super family of β-transforming growth factors. Teleost fishes do not have the Mullerian duct. Yoshinaga et al. (2004) have demonstrated for the first time the dimorphic expression of *amh* mRNA in *P. olivaceus*. The clone of *amh* of the flounder spans to 291 bp in length and the deduced amino acid sequence shows high identity to those of mammals and birds. *amh* mRNA has been detected in Sertoli cells surrounding spermatogonia but not in those surrounding spermatocytes and spermatozoa in the testis at 30 dph. It is also not detected in the ovary of the genetic and sex reversed phenotypic females. In the eel, a substance similar to *amh* labeled *eSrS21* is identified in Sertoli cells of immature testicles (Miura et al., 2002). Hence the regulation of spermatogonial proliferation and differentiation may be one of the common functions of *amh* in fish and amniote vertebrates, which possess the Mullerian ducts. There are also evidences for the role of *amh* in germ cell proliferation in *O. latipes*, being stimulatory during early embryonic stage but inhibitory at the onset of puberty (Shiraishi et al., 2008). In mammals, *amh* is glycosylated in the N terminal part of the protein region, which enhances the activity of the C terminal fragment. The C terminal fragment contains the conserved TGFβ domain. Antibodies raised against *amh* N and C terminal part of *amh* have been used to study the processing of endogenous and recombined *amh* of zebrafish. *amh* is processed to become bioactive and has independent functions in inhibiting both steroidgenesis and spermatogenesis.

(iv) Wilm's tumor suppressor gene (Wt1): Two *Wt1* genes have been identified in zebrafish (Perner et al., 2007) and medaka (Kluver et al., 2009). In medaka, it is concomitantly expressed in the somatic cells of gonadal primordium.

(v) *Sf-1*, the homologue of *Sf-1* of mammals, has been identified in medaka (Watanabe et al., 1999), salmon (Higa et al., 2000) and zebrafish (von

Hofsten et al., 2001). *Ff1b* and *Ff1d* are functional homologues of *Sf-1* identified in zebrafish; they show higher levels of expression in male gonads, especially in Leydig and Sertoli cells (von Hofsten et al., 2005a).

(vi) *cyp19* (*aromatase*): Despite the enormity of diversity in differentiation, the most conserved factor common to nearly all fishes is the control of ovarian differentiation by the gonadal aromatase gene *cyp19a1a*. The gene *cyp19* codes for the enzyme cytochrome p450 aromatase (p450 arom), a key enzyme in conversion of androgens to estrogens (Simpson et al., 1994). Aromatase is a member of the p450 cytochrome superfamily of enzymes. The role of aromatase and estrogens in sex differentiation in fish has recently been reviewed by Guiguen et al. (2010). The proportion of androgens and estrogens is critical for sex differentiation in many teleost species and is limited by the quantum of aromatase availability. p450 arom regulates the amount of estrogen, especially 17β-estrogen and maintains its level within adequate range for normal ovarian differentiation (Trant et al., 2001; Fenske and Segner, 2004). Aromatase is a duplicated gene in teleosts (Chiang et al., 2001), except in eels belonging to an ancient group (Cheshenko et al., 2008). The isoforms of the aromatase gene *cyp19a* and *cyp19b* encode two structurally different proteins p450 aromA and p450 aromB, respectively, with similar catalytic activities (Kwon et al., 2001a; Liu et al., 2007). p450 aromA and p450 aromB are preferably but not exclusively expressed in the gonads and brain, respectively (Piferrer and Blazquez, 2005). Restricted to gonad, *cyp19a1a* is directly involved in ovarian differentiation. Expressed in the brain, *cyp19a1b* acts indirectly probably through the hypothalamus-hypophysis-gonad axis (Chiang et al., 2001; Chang et al., 2005; Sawyer et al., 2006). However, Kallivretaki et al. (2007) have found that expression level of *cyp19a1b* during sex differentiation in zebrafish is not assigned to a particular sex. In the model proposed for sex differentiation in zebrafish by von Hofsten and Olsson (2005), p450 arom plays a key role.

p450 aromA is reported from the ovary and testis of many fishes but its enzyme activity is limited to the ovary only, although it occurs in *O. latipes* and *D. labrax* (Piferrer and Blazquez, 2005). In *O. niloticus*, a species with both genetic and environmental sex determination, high p450 aromA mRNA expression levels are required for ovarian differentiation (Kwon et al., 2001a,b). Two key steroidogenic aromatase enzymes known to be highly involved in gonadal differentiation are: (i) cytochrome p450 aromatase (p450 arom) and (ii) cytochrome p450 11β hydroxylase (p450 arom 11β, Kusakabe et al., 2002). In fishes, p450 11β is involved in the production of 11-oxygenated androgens, especially 11-KT, the potent androgen of males. The p450 11β cDNA has been cloned in *A. japonicus* (Jiang et al., 1996), *O. mykiss* (Liu et al., 2000) and *D. labrax* (Socorroa et al., 2007). In the trout, p450 11β is expressed specifically during testicular differentiation (Baron et al., 2005b). Androgens such as testosterone, DHT, or 11-KT have no direct effect on regulation of *cyp19a1b* expression in the brain of zebrafish (Mouriec et al., 2009).

(vii) *foxl2* is a *forkhead* box transcription factor (Hannenhalli and Kaestner, 2009) and is a female specific gene known to induce the earliest dimorphic sexual differentiation. It directly binds testis specific *sox9* enhancer element (TESCO) and silences the *sox9* expression in the ovaries (Uhlenhaut et al., 2009). Hence *fox12* is a suppressor of testes fate. It regulates the expression of *p450 arom* during sexual differentiation and signals the transcriptional regulation of *cyp19a1* (Yamaguchi et al., 2007). In *S. meridionalis, foxl2* is found to be expressed extensively in the brain and pituitary and with the highest level in the ovary, indicating the possible involvement of *foxl2* in the brain-pituitary-ovary axis (Liu et al., 2007). The association of *foxl2* SNP (Single Nucleotide) polymorphism in *P. olivaceus* has shown that the SNP1 in the forkhead domain is significantly associated with gonado somatic index (GSI), SNP2 downstream forkhead domain with serum 17β-estrodiol (E_2) level and SNP3 in the 3'- UTR with hepatosomatic index (HSI) (Shi et al., 2009).

(viii) *DaxL* is atypical of the family of nuclear receptors to hormones and is linked to chromosome XX in mammals (Zanaria et al., 1994). It has been isolated and characterized in many fishes, e.g., *O. mykiss* (Baron et al., 2005a), *Dicentrarchus labrax* (Martins et al., 2007) and *Squalius alburnoides* (Pala et al., 2009). It is known not to express in germ cells but in Sertoli cells and in follicular cells surrounding perinuclear oocytes in the adult female *S. alburnoides* (Pala et al., 2009).

(ix) *fig α*: is a germ cell specific transcription factor required for ovarian follicle formation. The expression of *fig α* and *zpe* coincides with the onset of gonadal differentiation in zebrafish (Jorgensen et al., 2008). Many genes, which are more known in the genetic cascade of mammals, may soon be identified and characterized in fishes. Some examples are the *figl α*, which are expressed in Sertoli cells of *S. alburnoides* (Pala et al., 2009), and *tbx1a*, a homologue of *tbx*, is highly expressed in somatic cells surrounding germ cells of the differentiating testis of *O. mykiss* (Yano et al., 2011). In fishes the existence of *shbga* and *shbgb* encoding different proteins are reported. *Shbga*, the mammalian homologue of *shbg*, found in almost all teleosts, is expressed in the liver. In contrast, *shbgb* is a salmonid-specific gene, and is expressed mainly in the ovary (Bobe et al., 2010).

2.4 Sex steroid receptor genes

From the point of endocrine distruption, sex steroid receptors are recognized but publications on them are few. Piferrer and Guiguen (2008) are the first to provide a brief account. The sex steroids exert their biological functions, when bound to specific receptors. The sex steroid receptors are the ligand-activated nuclear transactivation factors and control gene expression in all vertebrates. From the point of binding and molecular characterization, they are grouped into androgen (*ar*), estrogen (*es*) and progestogen (*ps*) receptors. Within *ar*, there are two types namely *arα* and *arβ*. Table 12 summarizes available information on identification, characterization and cloning of *ar* in a half dozon gonochoric fish. The *ar* mRNA level is shown to coincide with male differentiation in *Dicentrarchus labrax*. In zebrafish, the levels seem to determine the initiation of male differentiation period between 16 and 22 dph (Fig. 8).

Four piscine estrogene receptors (*er*) namely *erα*, *erβ*, *erβ2* and *erγ* have been identified (Hawkins et al., 2000). Their identification, characterization and expression in some fishes are briefly summarized in Table 13. *Er*s have high affinity for E_2 and specifically bind estrogen agonists like ethynylestradiol (EE_2) and diethylstilbosterol (DES) (Barrero-Gil et al., 2004)

Table 12. Summary of androgen receptors in gonochoric fish

Species	Observations	Reference
Oncorhynchus mykiss	*arα* fails to distinguish between any of the natural androgens, as tested by transactivation assay	Takeo and Yamashita (2000)
Anguilla anguilla	Identification of *arα*, *arβ* and *pr*	Ikeuchi et al. (2001)
Gambusia affinis	Two *ar* cDNAs are cloned and identified as *arα* and *arβ*. Both of them predominantly express in the distal anal fin rays	Ogino et al. (2004)
Dicentrarchus labrax	First assessment on the role of androgen receptor. Higher *ar* mRNA levels at 200 to 300 dph coincide with male differentiation	Blazguez and Piferrer (2004), Muriach et al. (2008)
Danio rerio	Molecular cloning and characterization of a nuclear *ar* mediates the action of 11-KT	Olsson et al. (2005)
D. rerio	The estimated *ar* levels determine the period of initiation of male differentiation among juvenile hermaphrodites	Jorgenssen et al. (2008)

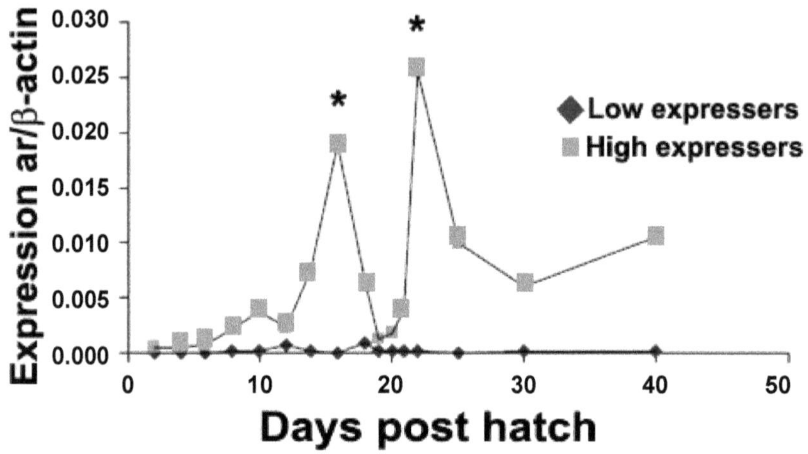

Fig. 8. Expression of androgen receptor in juvenile zebrafish during sex differentiation. Note the elevated expression at 16 and 22 dph among high expresser group, suggesting that the male differentiation is apparently initiated between 16 and 22 dph (from Jorgensen et al., 2008)

as well as some synthetic antagonists like ICI 182,780 (Piferrer and Guiguen, 2008). They produce several different transcripts in identified tissues in tested fishes. However, the evidence is limited, and not consistent, as some of them differ from species to species and sometimes even from tissue to tissue within a species. Future research may allow some generalizations.

Besides these nuclear steroid receptors, G-protein coupled membrane receptors have also been identified. Interestingly, they require just a few minutes and hours to elicit responses, and may play a role in the final maturation of gametes (Thomas et al., 2006). In the greenback flounder *Rhombeosolea tapirina* and rainbow trout *O. mykiss*, Hobby et al. (2000) have detected a sex Steroid Binding Protein (SBP) that binds E_2 with high affinity. Binding of E_2 to SBP is very rapid and occurs within 5 minutes. In trout, SBP binds E_2 with the highest affinity followed by Testosterone (T). In contrast, the relative affinity of T for flounder SBP is more than twice that of E_2.

In zebrafish, expression of mRNAs of gonadotropin receptors *fshr* in Leydig cells and *lhcgr* in Sertoli cells has been reported. The expression of *lhcgr* in Sertoli cells, the first report for a vertebrate, may be associated with the undifferentiated gonochoric mode of gonadal differentiation in zebrafish. In the recombinant zebrafish (*rzf*), *rzfLH* and *rzfFSH* stimulate the release of gonadotropins (Garcia–Lopez et al., 2010).

Incidentally, the evolutionary history of receptor genes seems to provide an explanation for the indecisive or plasticity of sex determination in fishes. Two Whole Genome Duplication (WGD) events are considered to have occurred before the tetrapod split from the fish 360 MYa (Zhou et al., 2002). In an interesting contribution, Douard et al. (2008) have described

Table 13. Summary of estrogen receptors in gonochoric fish

Species	Observations	Reference
Danio rerio	Molecular characterization of *erα*, *erβ*$_1$ and *erβ*$_2$. Their binding characteristics, transactivation and expression in selected tissues	Menuet et al. (2002)
Micropogonias undulatus	Unusual binding properties of *erβ* are accompanied by highly conserved amino acid changes in the ligand binding domain	Hawkins and Thomas (2004)
Pimephales promelas	Molecular characterization of *erα*, *erβ*$_1$ and *erβ*$_2$. Ontogenic expression profile in selected tissues	Filby and Tyler (2005)
Fundulus heteroclitus	Cloning of *erα*, *erβ*$_1$ and *erβ*$_2$. Difference in levels of expression in populations exposed to pollutants	Greytak and Callard (2007)
Dicentrarchus labrax	Characterization, cloning and expression of *erα*, *erβ*$_1$ and *erβ*$_2$. Expression of *erα* in pituitary and gonad at 200 dph but highest expression of *erβ*$_1$ and *erβ*$_2$ in gonads at 250 dph	Halm et al. (2004)
Oncorhynchus mykiss	Identified *erα*$_2$ and isoforms of *erβ*	Nagler et al. (2007)
Paralichthys olivaceus	Successful development of an *erα* and *erβ*-mediated reporter gene assay using Vg promoter as an estrogen responsive promoter	Kitano et al. (2006)
P. olivaceus	Tamoxifen-treated genetic females suppress *cyp19a* and induce *amh*, but do not affect *erα* and *erβ*	Kitano et al. (2007)
Oreochromis niloticus	Ontogenic expression patterns of *ers*; *erα* expression earlier than *erβ*. Notably, the appearance of *erα* and *erβ* is earlier than steroidogenic enzymes	Sudhakumari et al. (2005)

the duplication and fate of the androgen receptor (*ar*), a nuclear receptor known to play a key role in sex determination in vertebrates. Consistent with an early WGD event, *ar* has been duplicated into two genes *ar-A* and *ar-B* (e.g., *Anguilla, Heterotis*) after the split of Acipenseriformes from the lineage leading to teleost fishes but prior to the divergence of Osteoglossiformes. Genomic and syntenic analyses have shown that one of the duplicated copies, *ar-B* has been lost in several Clupeocephalids such as Cypriniformes (five species including zebrafish), Characiformes (one species), Siluriformes (one species) and Salmoniformes (two species) but not the basal Osteoglossiformes and Anguilliformes. In Percomorphs, *ar-B* has accumulated substitutions in the ligand-binding and DNA-binding domains (LBD and DBD).

Further analysis by Douard et al. (2008) has led them to suggest the putatitive neo-functionalization of the same duplicate in Percomorphs, which occurred a long time after the WGD. For instance, *ar-A* and *ar-B* have distinct expression patterns in the brain of *Astatotilapia burtoni* (Harbott et al., 2007) with differential amplification of these receptors in the maintenance social dominance status of the male (Baumeister et al., 2007). Considering such functional shifts observed in the Percomorph lineages, Douard et al. (2008) have been tempted to link their functional shift and sexual lability, in view of the fact that the Percomorphs contain 90% of all hermaphrodite species known to date. The existence of two functionally divergent *ar-A* genes may be viewed as a permissive factor allowing plasticity (cf Le Page et al., 2010) and evolvability of divergent sex determination in these fishes.

2.5 Gonadal differentiation

In most teleosts, ovarian differentiation precedes and commences with the appearance of proliferation of germ cells supporting somatic cells, arrival of germ cells, followed by meiosis, and the formation of oogonia, early oocytes and ovarian cavity (Guraya, 2000). For instance, the proliferation of PGCs in embryos of chub mackerel *Scomber japonicus* is completed between 5 and 10 dph. The gonadal primordium is formed at 15 dph, followed by ovarian differentiation between 30 and 40 dph, as indicated by the presence of meiotic oocytes and the ovarian cavity. Perinuclear oocytes are observed at 60 dph (Kobayashi et al., 2011). Due to delayed proliferation of germ cells, both ovarian and testicular differentiation commences simultaneously in *Siganus guttatus* (Komatsu et al., 2006), which is in contrast to the general norm of teleosts. However, testicular differentiation in *S. japonicus,* initiated at 30 dph, as indicated by the formation of sperm duct primordial, develops into spermatocytes at the chromatin nuclear stage after 90 dph. Subsequently spermatids and spermatozoa appear at 160 dph (Kobayashi et al., 2011).

Figure 21 summarizes a generalized chronological sequence of gonadal differentiation events in a teleost.

From relevant literature, Saito and Tanaka (2009) summarize a detailed account on gonadal differentiation in the Japanese medaka *O. latipes* and the same has been chosen as a model to present an overview. Manifestation of Sertoli cells in males and granulosa cells in females is considered as the 'starting point' for testicular and ovarian differentiation, respectively. Sertoli cells are known to orchestrate development and morphogenesis of testis, which then signals the sexual fate of the entire reproductive system through production of hormones (Brennan and Capel, 2004).

Gonads: Gonadal development commences with the formation of sexually undifferentiated primordium consisting of the germ cells namely Primordial Germ Cells (PGCs) (Fig. 13) and supporting somatic cells. Differentiated at the early blastula stage, PGCs migrate to colonize the gonadal anlage. Then the PGC clusters are isolated from each other by one of the somatic gonadal precursors expressing *sox9b*. Subsequently, a single gonadal primordium consisting of the PGCs and somatic cells is separated into right and left lobes. On the 4–5th day post-fertilization (dpf), a leaf-like gonadal primordium is suspended at the mid-dorsal mysentery with protruding right and left lobes (Hamaguchi, 1982). The number of PGCs can be experimentally manipulated by disturbing the migration of PGCs to the gonadal primordium (see also Table 22), which provides a useful sex reversal model for understanding the role of PGCs in gonadal differentiation (Saito and Tanaka, 2009).

Ovary: Following the two to four consecutive rounds of divisions, many cysts each consisting of four, eight or 16 cells are formed in the female embryo at 4–5 dpf (Saito et al., 2007). Hence the number of PGCs is larger at hatching in females than in males, which is the first known sign of manifestation of ovarian differentiation. Subsequently, the germ cells enter meiosis synchronously. Oogenesis persists throughout the gonadal development and in the mature ovary (see Fig. 32).

The earliest manifestation of somatic cell differentiation is induced by the female specific expression of *aromatase* gene at 1–2 weeks post-fertilization (wpf) in the cluster of somatic cells located at ventral area of developing gonad (Suzuki et al., 2004). These cells migrate dorsally and surround the developing oocytes. They become morphologically recognizable by 3–4 wpf. They are the precursors of the ovarian follicular granulosa cells, as detected by *foxl2*, a granulosa marker gene (Nakamura S et al., 2008).

Testis: Genes, responsible for the earliest induction of gonadal differentiation, have been identified as *DMY/Dmrt1bY* in *O. latipes*, *tDmrt1* in *Oreochromis niloticus* (see Fig. 13), and *amh* in *Danio rerio*. At the induction of one or the other of these genes, the proliferation of PGCs is postponed for a while in the presumptive male embryos, resulting in the production of significantly less number of PGCs in male embryos than in female embryos in the three species, for which the relevant information is reported (Fig. 13). This is perhaps the first sign of testicular differentiation. The expression of these genes is localized in some germ cell supporting somatic cells (GCSSCs) that surround the PGCs. They are the pre-Sertoli cells and differentiate into Sertoli Cells (Kobayashi et al., 2004).

In *O. latipes* and *O. curvinotus Casp6* is a sex linked DNA marker (Kondo et al., 2003). Males have *DMY* and are heterogametic in *Casp6*, whereas females are homogametic in *Casp6* (Matsuda et al., 2003). Males carry paternal genotype and females maternal genotype, indicating the *DMY* is located on the Y chromosome. The cDNA sequence of *DMY* of *O. latipes* encodes a putative protein of 280 amino acids, whereas the cDNA sequence of *Dmrt1* a protein of 276 amino acid (Fig. 9B). Many common PCR primers can be used on *DMY* and *Dmrt1*, as they carry similar nucleotide sequence. A PCR primer *DMY* and *Dmrt1* of *O. latipes* amplifies *DMY* and *Dmrt1* of *O. curvinotus* (Fig. 9A).

A phylogenetic tree based on amino acid sequences of the *DM*-domains of these and other *DMY* and *Dmrt1* in database shows that *Oryzias DMY* makes a clade with *Oryzias Dmrt1*. The clade is divided into two lineages, one consisting of the *DMYs* of *O. latipes* and *O. curvinotus*, and the other with the *Dmrt1s* of these two species (Fig. 9C). Hence *DMY* has been derived from *Dmrt1* immediately prior to separation of *O. latipes* and *O. curvinotus* (Matsuda et al., 2003).

Age of *DMY*: *DMY* is located on the Y chromosome, which is always present in males of *O. latipes*, while *Dmrt1* located on an autosome, is present both in males and females. The hypothesis of male-driven evolution asserts that the mutation rate is higher in males than in females due to more rounds of cell divisions in the male germ cells (Li et al., 2002; see also Pandian, 2011). Since the duplication events (see Pandian, 2011), there are a total of 31.6 synonymous mutations in *DMY* but only 18.3 in *Dmrt1* branches. Considering the nucleotide mutation rate in mammals (Kumar and Subramanian, 2002), the duplication occurred 15.3 Mya in *Oryzias*. Taking synonymous substitution rates in other fishes into consideration, Zhang (2004) estimated that the age of *DMY* is in order of 10 My.

Sex determination involves a complex hierarchy of genes. Expression screen analysis has identified hundreds of candidate genes that show sex specific expression pattern. It has indeed been difficult to place these genes and their regulatory function into a cascade/network of genes. Multiple Transposable Elements (TEs) are known to have been inserted into the Y-specific region on LG1 of medaka (Kondo et al., 2006). Recently

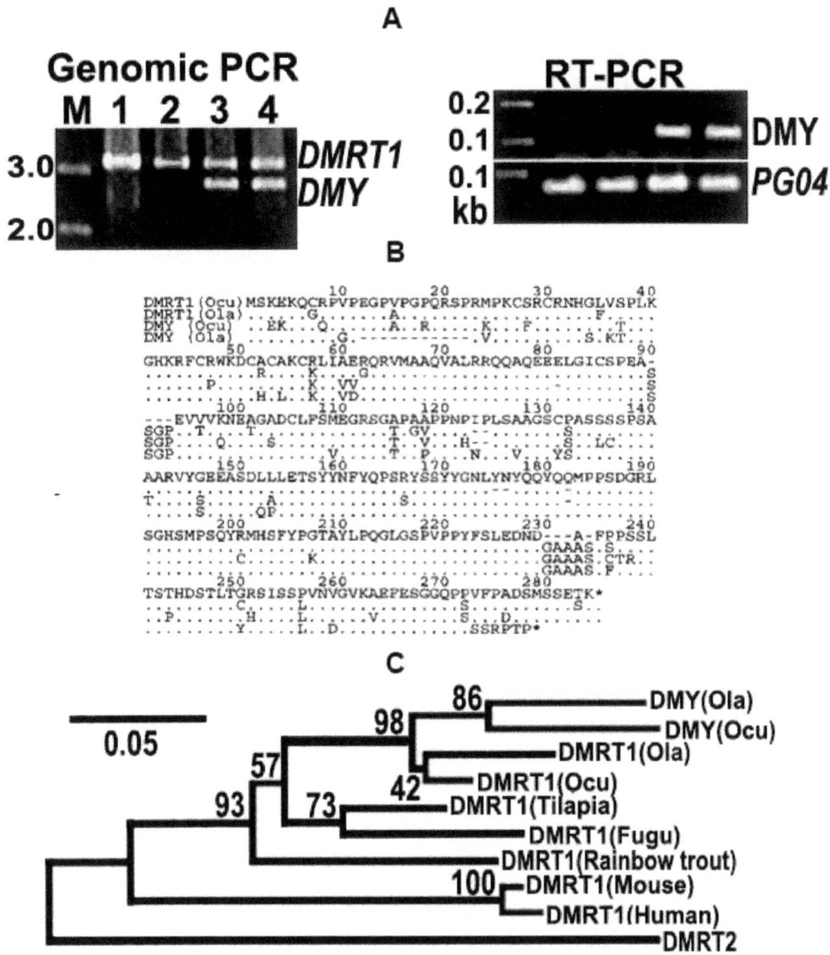

Fig. 9. Upper panel (A): The *Dmrt1* and *Dmy* of *Oryzias latipes* are amplified by the *Dmrt1* and *Dmy* of *O. curvinotus* Middle panel (B): Alignments of Dmy and Dmrt1 proteins of *O. latipes* (Ola) and *O. curvinotus* (Ocu) period indicated the sequence identity: dash a gap and asterix, a stop codon. A single base pair deletion at position 272 of Ola *Dmy* has caused a frame-shift mutation. Lower panel (C): A neighbor-joining tree based on the amino acid sequences from positions 21 to 109 in the middle panel (from Matsuda et al., 2003; Zoological Science, 20: 159–161)

Herpin et al. (2010) have found that the *cis*-regulatory element containing the *Dmrt1*binding site, pre-existing within *zanagi* element at the time of its insertion, was co-opted to confer *Dmrt1bY* its expression pattern after gene duplication. This finding reveals an important role for TEs in transcriptional network rewiring and initiating a new function in the duplicated genes and thereby creating new hierarchies of sex determining genes. Briefly, despite the commonality of the end point, there seems to be endless ways, in which the sex chromosomes are differentiated.

Evolution of *DMY*: However, *DMY* is not found in *O. celebensis, O. mekongnensis, P. reticulata, O. niloticus, D. rerio* and *Fugu rubripes* (Kondo et al., 2003; Tanaka et al., 2007). Notably, *O. latipes* and its sister species *O. curvinotus* and *O. luzonensis* possess an XX/XY sex determination system. However, the sex chromosomes of *O. luzonensis* display only a weak reduction of the recombination frequency and recombine well around the sex determining region, in contrast to the highly restricted sex chromosome of *O. latipes*. Hence the sex chromosomes of *O. luzonensis* must be 'younger' than that of *O. latipes* and may be of 5 My (Tanaka et al., 2007). Similarly, the Y chromosome in *O. dancena* is not homologous to that in *O. latipes,* although both of them have an XX/XY sex determination system. Through progeny testing and linkage analysis of isolated sex-linked DNA markers, Takehana et al. (2007) have shown that *O. hubbsi*, a most closely related species to *O. dancena* has a ZZ/ZW system. Further the sex chromosome of *O. hubbsi* and *O. dancena* are not homologous, indicating different origins of these sex chromosomes systems.

Mutations and sex reversal: Despite the estimated long history of *DMY* over 10 My, there are evidences to show that the Sex Determination System (SDS) has not progressed to a similar level of stability, as in mammals and birds. Firstly, the presence of homomorphic X and Y chromosomes in medaka fish indicates that the molecular differentiation process of the sex chromosomes has not progressed to a level of complete recombinational isolation to render the visibility of X and Y chromosomes. Secondly, the sex specific region in the Y chromosome is too small (cf Woram et al., 2003), spanning to a few hundred kilobase in length. Some examples are described to show that the SDS in medaka fish remains unstable rendering the spontaneous appearences of XX males and XY females.

PCR genotyping of various strains of medaka, Nanda et al. (2003) found a strikingly large number of males (> 10%) that did not amplify *Dmrt1bY* (Fig. 10A,B); this was also confirmed by Southern blot analysis (Fig. 10C). All these aberrant males had only X chromosomal alleles (Fig. 10D,E). Incidentally, as early as in 1936, Aida recorded that of 5,000 medaka males analyzed, seven were detected as XX males in one laboratory strain.

Shinomiya et al. (2004) surveyed 2,274 wild-caught *O. latipes* from 40 localities spread over all of Japan and 730 fishes from 69 wild stocks from Japan, Korea, China and Taiwan, and identified 23 XY females from 11 localities (Table 14). Genetic analysis revealed that the XY females from eight localities produced all female XY^{m}, XY^{Sai} or XY^{wOur} progenies (carrying $Y^{m}/Y^{Sai}/Y^{wOur}$ chromosome derived from XY females), while those from the

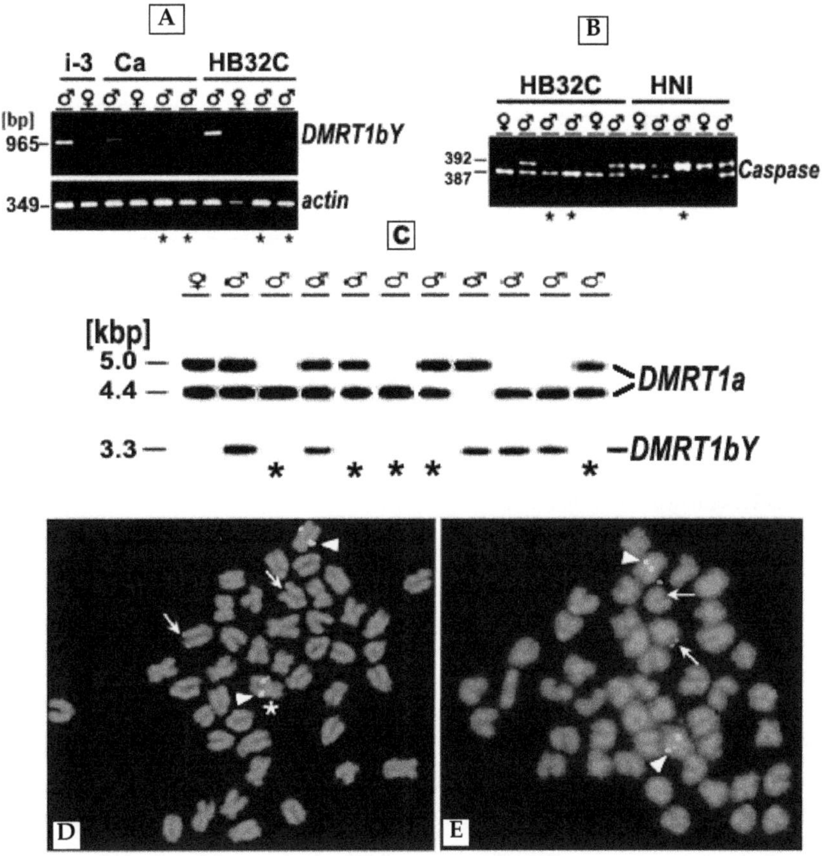

Fig. 10. *Oryzias latipes*: (A): PCR genotyping of males and females. Note the presence of *Dmrt1bY* in normal males but its absence in aberrant females. (B): Hemizygosity of normal males of the HB32C and HNI strains for *caspase* 6. Females and aberrant males (*) show only the X chromosomal PCR product. (C): Southern blot analysis of DNA from female, normal male and aberrant (*Dmrt1bY* negative) males (*). FISH pattern of Y specific and sex chromosome specific probes on (D) XY and (E) XX metaphase chromosomes. Note the presence of three hybridization signals in XY males, as compared to two spots in XX males (from Nanda et al., 2003, with permission by the Genetic Society of America).

Color image of this figure appears in the color plate section at the end of the book.

other three localities yielded both male and female XY^m progenies. Hence all these wild XY females had Y-linked mutations.

Screening for Y-linked mutations in amino acid coding sequence of *DMY* (Fig. 11) and examining *DMY* expressions at 0 dph with densitometric semi-quantitative RT-PCR, Otake et al. (2006) classified the mutations into two groups: one containing mutations in the amino acid coding sequence and the other showing reduced *DMY* expression (cf Herpin et al., 2007), although

Table 14. List and functions of mutant XY female from wild medaka populations of Japan (compiled from Otake et al., 2006, with permission by the Genetic Society of America)

Site	Prefecture	Genotype	Phenotype sex of XY^mF_1	Remarks
Aomori	Aomori	XY^m	all ♀♀	
Aizu-bange	Fukushinia	XY^m	all ♀♀	
Kurobe	Toyama	XY^m	all ♀♀	Mutant *DMY* gene is not functional
Suzu	Ishikawa	XY^m	all ♀♀	
Awara	Fukui	XY^m	all ♀♀	
Aki	Kochi	XY^m	all ♀♀	
Oura	Kagoshima	XY^{wOur}	all ♀♀	Lost ♂ determining function
Saigo	Shimane	XY^{mSai}	all ♀♀	
Kesen-numa	Miyagi	XY^{wKsn}	♀♀ + ♂♂	Two XY^{wKsn}/XY^m are required to determine ♂ sex
Auza-Wakamatsu	Fukushima	XY^m	♀♀ + ♂♂	
Shirone	Niigata	XY^m	♀♀ + ♂♂	

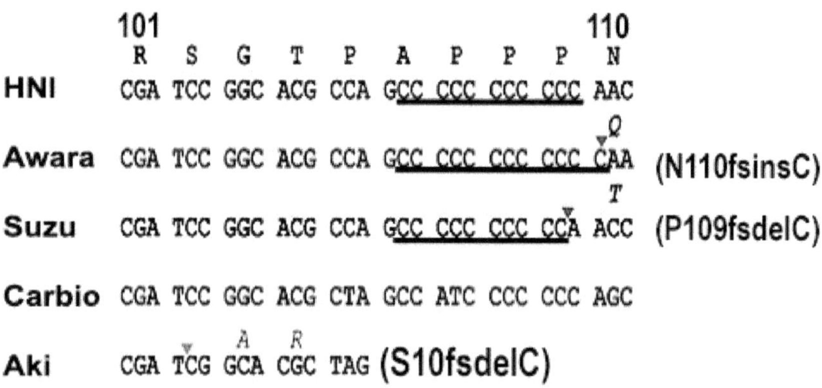

Fig. 11. Nucleotide sequences sex reversed mutant and medaka *Oryzias latipes*. The poly C tract specific to northern populations of HNI, Awara and Suzu are underlined. The Awara XY ♀ has a cytosine insertion at amino acid 110, the Suzu XY♀ cytosine deletion at amino acid 109 and the Aki XY ♀ cytosine insertion at amino acid 102, each of which results in a frameshift and premature termination. The nucleotide sequences of Hd-rRY HNI, and carbio XY females are mutants at the poly C tract in exon 3 (from Otake et al., 2006, with permission by the Genetic Society of America)

with normal *DMY* coding sequence (Fig. 11). The mutations included (i) a C insertion in a poly C tract and deletions in exon 3, (ii) a C deletion in the same poly C tract in exon 3, (iii) a C deletion in another site in exon 3 and (iv) a large insertion or deletion in intron 5 and/or exon 6 in the Saigo strain ($Y^{mSai}Y^{mSai}$) XY females and possess no polyadenylation signal.

Hd-rR^{HNI} male sired male and female progenies, when mated to Kesen-numa XY ^{wKsn} or Shirone XY^{m} females. Interestingly, four females among 18 XY^{wKsn} F_1 progenies and 22 XY^{wKsn} females among 35 F_2 progenies were produced. Conversely, all the 41 F_2 progenies (bearing Y^{wKsn}/Y^{wKsn} genotype) produced from a cross between F_1 XY^{wKsn} ♀ and F_1 Y^{HNI}Y^{wKsn} ♂ were all males. These observations on sex ratio of F_1 and F_2 progenies indicate that a minimum of two Y^{wKsn} chromosomes are required for male sex determination, a phenomenon resembling the female heterogametic system (Matsuda, 2011). All progenies sired by Hd-rRY^{HNI} male mated to XY^{wOur} female were females (Otake et al., 2006).

Otake et al. (2006) also measured the *DMY* expression levels in some of these XY females. The *DMY* mRNA levels are significantly lower in Y^{wKsn} Y^{wKsn} than in XY$^{Hd\text{-}rR}$ and XY^{HNI} males (Fig. 12). The same is also true of Y^{wOur} Y^{wOur}. All these progenies carrying XY^{wKsn}XY^{wKsn} and Y^{wOur}Y^{wOur} genotypes develop into females.

2.6 Model fishes

Of four model species, in which sex differentiation is described, almost complete, detailed information is available for *O. latipes* thanks to Japanese scientists. For the others, information available is limited and fragmentary in the order of *O. niloticus* < *D. rerio* < *Silurus meridionalis*. Yet these models provide interesting diversity in sex differentiation and represent the first step in our endeavour to identify the genes responsible for induction of sexual dimorphism; the very first gene known at present to induce male differentiation is *DMY/Dmrt1bY* in medaka, *tDmrt1* in tilapia and *amh* in zebrafish; in the Southern catfish, *foxl2* may be the first to induce female differentiation, as the ovarian cavity is formed prior to the proliferation of PGCs.

2.6a *Oryzias latipes*

In medaka fish, the formation of gonadal anlage is completed at the embryonic stage 38 (Kobayashi et al., 2004). The very first sign of sexually different trends in proliferation of PGCs becomes apparent at stages between 37 and 39, when *DMY* is specifically expressed in male XY embryos but not in XX embryos (see also Lewis et al., 2008). Incidentally, the *Dmrt1* mRNA is not yet detectable at these stages clearly indicating that *DMY* begins its expression much earlier than *Dmrt1* and any morphological differentiation such as the formation of acinus, a globular structure and the precursor of the seminiferous tubules (see Pandian, 2010).

Figure 13 shows bisexually different trends of proliferation of PGCs in selected fishes. The first role of *Dmrt1bY* is the suppression of proliferation of PGCs through a paracrine system. The proliferation is even inhibited

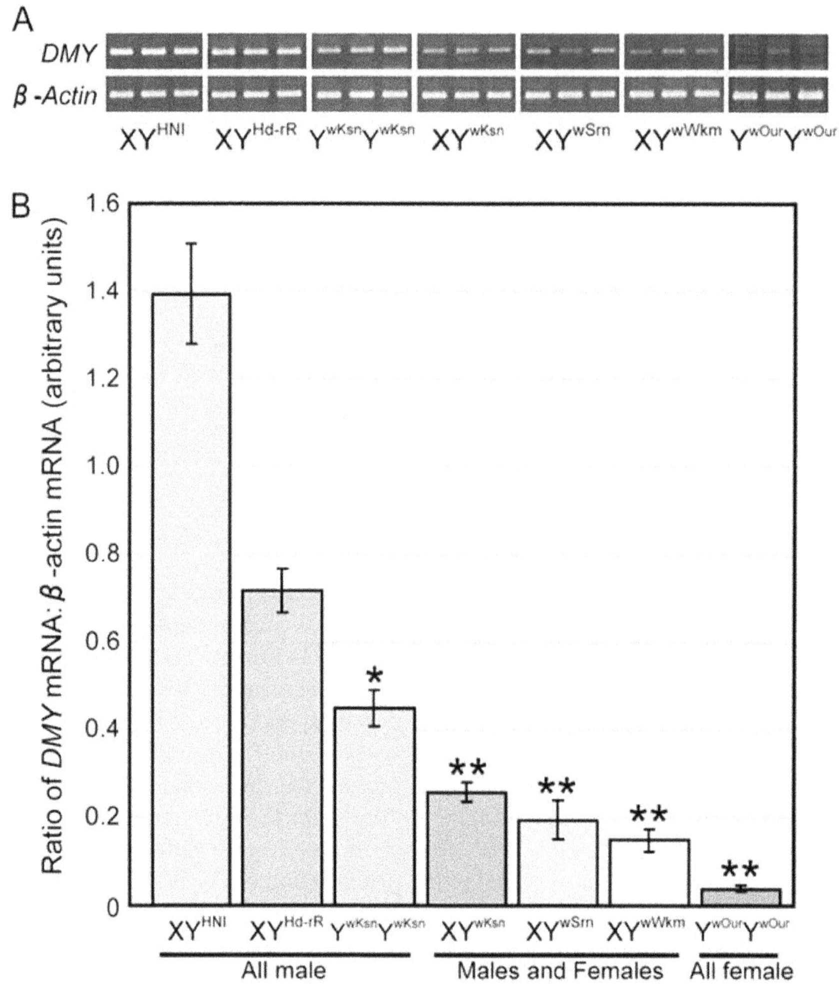

Fig. 12. Expression levels of *Dmy* transcripts at 0 dah in XY^{HNI}, $XY^{Hd\text{-}rR}$ and sex reversed mutants of *Oryzias latipes.* (A) Ethidium bromide-staining of the *Dmy* transcripts following gel electrophoresis. β-actin is loaded as a loading control: (B) Columns show the ratio of *Dmy* mRNA β-actin mRNA in arbitrary unit. Astriks indicate statistically significant differences between the mean values of XY^{HNI} and $XY^{Hd\text{-}rR}$ as well as XY^{HNI} $XY^{Hd\text{-}rR}$ and Y^{wKsn} Y^{wKsn} (from Otake et al., 2006, with permission by the Genetic Society of America)

in embryos overexpressing *dmrt1bY* and cells are arrested in the G2 phase of the cell cycle (Herpin et al., 2007). Hence, the proliferation in male embryos commenced later than in female embryos, when *Dmrt1bY* acts in an autocrine manner (Kondo et al., 2009). The proliferation of PGCs is accelerated from stage 37 in XX embryos and reaches a peak with about 125–225 cells at stage 39. However, in XY embryos, after a brief period of

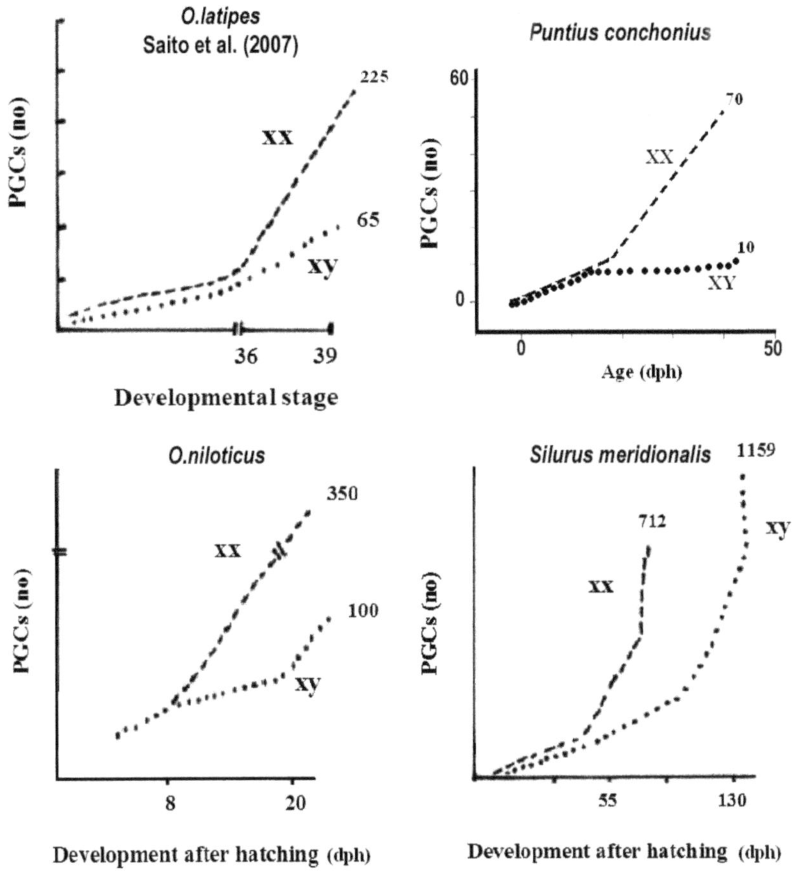

Fig. 13. Schematic representation of PGC proliferation as function of developmental stage/ days after hatching in selected model fishes: Upper panels: Left, *Oryzias latipes*, Right, *Puntius conchonius* Lower left panel: *Oreochromis niloticus* and Lower right panel: *Silurus meridionalis* (compiled from Saito et al. (2007), Kobayashi et al. (2004, 2008) Cek et al. (1998) and Liu et al. (2008))

arrest, it progresses slowly to accumulate only about 65 and odd cells at stage 39 (Kobayashi et al., 2004). Saito et al. (2007) have more elaborately described the proliferation of PGCs in XX embryos, and the proliferation is continued with intermittent divisions producing isolated daughter cells. Two to four rounds of continuous divisions form cysts each with four, eight and 16 cells, which subsequently enter meiosis synchronously. Interestingly, the *zenzai* mutant medaka, a defect of slow intermittent division, eventually leads to the depletion of PGCs in adult gonads of both sexes, despite the cyst formation process remaining unaffected. Conversely, excessive proliferation of PGCs occurs in the mutant *hotei* with Y3qoe mutation in *amh* receptor type

II (Morinaga et al., 2004, 2007). This *hotei* medaka possesses an expanded abdomen with hypertrophic gonads. Interestingly functional male-to-female sex reversal frequently occurs in *hotei* mutant genetic male and 50% of *hotei* homozygous males are sex reversed. Surprisingly, the *amh* is also involved in sex-dependent regulation of PGC proliferation (Shiraishi et al., 2008).

In the normal XY embryos of medaka, *DMY* continues to express on the zero dah and its 36 KD protein is present primarily in the nuclei of PGC supporting somatic cells surrounding the PGCs. But after 20 dph, the *Dmrt1* mRNA begins to increase more rapidly in testes than in the XX gonad. Subsequently, the *Dmrt1* mRNA is more rapidly accumulated than that of *DMY* mRNA. Expression of *DMY* persists in Sertoli cells lineage from PGC-supporting cells upto Sertoli cells, indicating that only *DMY*- positive cells enclose PGCs. Hence the *DMY*-positive cells are the pre-Sertoli cells and differentiate into Sertoli cells. In mature testes, *DMY* is localized in the Sertoli cells during spermatogenesis and epithelial cells of intratesticular efferent duct, a derivative of Sertoli cells. *Dmrt1* is expressed in spermatogonium supporting cells after testicular differentiation and its expression is much higher than that of *DMY* in mature testes. In testes of sex reversed (XX) neomales, which never expresses *DMY* (as XX ♀ gonad), *Dmrt1* mRNA is localized in Sertoli cells and the epithelial cells of intratesticular efferent duct, clearly showing that *DMY* and *Dmrt1* are co-expressed in Sertoli cells in mature testes. This important observation of Kobayashi et al. (2004) in *O. latipes* has much implication on the role played by *Dmrt1* in gonadal differentiation of *O. niloticus*. Briefly *DMY* regulates proliferation of PGCs and sex specific early embryonic differentiation, whereas *Dmrt1* regulates spermatogonial differentiation during subsequent embryonic gonadal differentiation.

In medaka, Yokoi et al. (2002) isolated two kinds of transcripts *sox9* and *soxlf*. The *sox9* encodes 487 amino acids and shows 70% amino acid sequence identity with known vertebrate *sox9* proteins. *soxlf* is a longer form of the *sox9*, which transcribes from an additional exon in 5′ upstream region. Nakamoto et al. (2005) cloned a novel testicular type *sox9a* from medaka, which they named *sox9a2*. Establishing an enhanced green fluorescent protein *Egfp* transgenic line that mimics *sox9b/sox9a2*, Nakamura S et al. (2008) found a cluster of *Egfp*-expressing cells in single gonadal primordium. In this primordium, they also characterized the precursors of the supporting somatic cell lineage common to both Sertoli cells and granulosa cells in embryos at stage 3. In an earlier study, Nakamura et al. (2006) showed expression of *sox9b/sox9a2* gene, a homologue of mammalian *Sox9*, in one of the supporting somatic cell line of the medaka; in fact two distinct somatic supporting cell lineages, one with *sox9b/sox9a2* expressing cells and the other with *ftz-f1* expressing cells are formed prior to the formation of the gonadal primordium. The fate of these cell lineages, however, has not yet been fully elucidated.

foxl2 is involved in ovarian development and in medaka, it commences its expression in PGC-supporting somatic cells in XX gonad, just after the initiation of ovarian differentiation. It is maintained in granulosa cells throughout ovarian development and is involved in differentiation of granulosa cells. In the adult ovary, *foxl2* is expressed among pre-vitellogenic and vitellogenic follicles but not in post-vitellogenic follicles (however see Nakamoto et al., 2006).

Recording differences in the frequency of spontaneous occurrence of male progenies among XX♀ x XX♂ breeding pairs of *O. latipes,* Scholz et al. (2003) have found that the *Dmrt1bY* expression is strictly correlated with the genotype but not with phenotype, i.e., its expression is restricted to XY and YY gonads, regardless of the phenotypic sex. In contrast, the expression of *fig1a* is correlated with the phenotypic sex. *Dmrt1bY* seems to play a key regulatory role at an early event of sex determination and remains hormone-insensitive, as against the hormone-sensitive *tDmrt1* in *O. niloticus* (Kobayashi et al., 2008).

More recently, a new factor named gonadal-soma-derived factor (*Gsdf*) has been found to be associated with testicular differentiation (Sawatari et al., 2007). There are two forms *Gsdf* in rainbow trout *Gsdf1* and *Gsdf2,* while only one in medaka. A rise in expression levels of *Gsdf* in XY gonads around the 6th dpf (see Paul-Prasanth et al., 2011) leading to testicular differentiation. Whereas the expression of *Dmy* is linked to the gonotype of the gonads, that of *Gsdf* is closely linked to the gonadal phenotype.

2.6b *Oreochromis niloticus*

In the Nile tilapia, males are heterogametic (XY), as in medaka. The sequence of sexual dimorphic differentiation is similar to that of medaka but the position of male inducing gene *DMY/Dmrt1bY* in medaka is assumed by *Dmrt1* called *tDmrt1* in tilapia. Kobayashi et al. (2008) have shown that *tDmrt1* expression occurs specifically in Sertoli cell lineage and the expression precedes testicular differentiation. Examining the sexually dimorphic proliferation of PGCs and the expression profiles of *tDmrt1,* they have detected strong expression in supporting somatic cells surrounding the PGCs at 5 dph fry, prior to the sexually dimorphic proliferation numbers of PGCs and histogenesis such as the differentiation of intratesticular efferent duct or ovarian cavity. The number of PGCs, which are around 30 at 5 dph female fry, begins to progressively increase to about 350 at 20 dph fry. During the period from 5 dph to 20 dph, the expression of *tDmrt1* becomes progressively stronger. But the PGC number in the XY fry begins to increase to about 45 in an 8 dph fry and after a period of arrest, it peaks to about 100 at the 20 dph fry (Fig. 13). To have a better insight into the role played by *tDmrt1* 'as superior molecular marker' of inducing male differentiation in tilapia, Kobayashi et al. (2008) have treated XX tilapia fry with 17α-methyltestosterone (MT) from 6 to 14 dph. The mean number of PGC cells in the MT-treated XX fry is 98 and is less than that (135) in the normal XY fry. Hence the MT dose-dependent sex reversal is accompanied by sex specific *tDmrt1* expression. Incidentally, similar trends of faster and slower proliferation of PGCs in female and male embryos of the cyprinid rosy barb *Puntius conchonius* have been reported by Cek et al. (1998).

In the untreated normal Nile tilapia fry, the *tDmrt1* signals become localized in Sertoli cells and epithelial cells of intratesticular efferent duct, recalling similar events in the *DMY* induction in *O. latipes*. When the formation of the intratesticular efferent duct anlage is apparent at 25 dph (Fig. 14), the *tDmrt1* expression is detected in Sertoli cells, medullary mass and its derivative the intratesticular epithelial cells of efferent duct. Using laminin antibody, *tDmrt1* positive cells have been found only within the

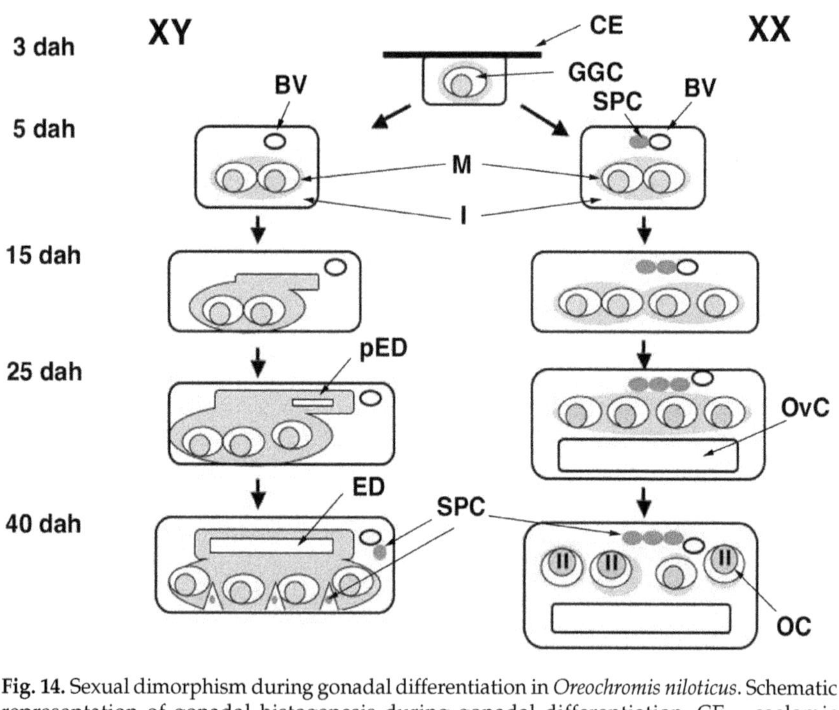

Fig. 14. Sexual dimorphism during gonadal differentiation in *Oreochromis niloticus*. Schematic representation of gonadal histogenesis during gonadal differentiation. CE = coelomic epithelium, GGC = gonial germ cell, SPC = steroid producing cell, BV = blood vessels, OVC = ovarian cavity, pED = anlage of intratesticular efferent duct, Oc = oocyte, ED = intratesticular efferent duct, M = medullary cell, I = interstitium (from Kobayashi, 2010, reproduced with permission by The International Journal of Developmental Biology, 54: 105–111)

seminiferous tubules, indicating that *tDmrt1* is specifically expressed in Sertoli cells lineage only.

The expression of *sox9a* has been detected in the PGC supporting somatic cells but no difference between sexes has been noted until the age of 25 dph in the fry. In contrast to *tDmrt1*, *sox9a* is most expressed in epithelial cells of efferent testicular duct in testes. RT-PCR and *in situ hybridization* studies have conclusively shown that unlike *tDmrt1*, the *sox9a* is expressed in XY gonads only after the appearance of intratesticular efferent duct or formation of ovarian cavity (Fig. 14). After the induction of XY embryo and followed by sex reversal with estrogen, the expression of *tDmrt1* is suppressed (by down-regulation) and then disappeared completely. In contrast, *tDmrt1* is expressed by up-regulation in the surrounding cells of PGCs in XX reversal with androgen (Kobayashi et al., 2008).

Kwon et al. (2001a) have cloned the brain *aromatase* gene and studied the expression of brain aromatase and ovarian aromatases. Their semi-quantitative RT-PCR analysis reveals that the expression of brain and ovarian *aromatase* genes is initiated between 3 and 4 dpf (day post-fertilization). The level of brain aromatase mRNA gradually increases throughout the period upto 27 dpf with little difference between sexes. A marked sexual dimorphism of ovarian aromatase mRNA becomes apparent between 15 and 27 dpf, i.e., the level is gradually up-regulated in females, while that in males is dramatically down-regulated. Incidentally the treatment with oral feeding of aromatase inhibitor (ATD, 1, 4, 6 androstatriene 3–17 dione) from 11 to 17 dpf or immersion of 13 dpf-fry in a solution containing aromatase inhibitor significantly increases the percentage of male tilapias (Kwon et al., 2000). These findings clearly indicate that the ovarian *aromatase* gene plays a decisive role in initiation of sexual dimorphism in *O. niloticus*.

A study on 17 sex differentiation genes in *O. niloticus* fry (5–15 dph and 25–70 dph) by Ijiri et al. (2008) has not only confirmed the described observations but also indicated the role of other sex differentiation genes. The expression of aromatase (*cyp19a1a*), an enzyme responsible for producing 17β-estrodiol, has commenced at 5 dah with marked elevation thereafter in XX gonads. The close relationship of *foxl2* in XX gonads suggests an important role in the transcriptional regulation of *cyp19a1a*. In contrast, *tDmrt1* shows male specific expression from 6 dah onward, suggesting its decisively important role in testicular differentiation. *sox9* and *amh* showed testis specific expression subsequently. From 35 dph only the mRNA expression of 11α-hydroxylase (*cyp11b2*), an enzyme responsible for the synthesis of a potent 11-Ketotestosterone (11-KT), has been found in XY gonads.

In *Oncorhynchus* too, the *aromatase* gene expression is observed at 55 (day post fertilization) dpf (= 20 dah) before the first occurrence of meiosis in the early oocytes (70 dpf = 35 dah) (Guiguen et al., 1999). The level of its expression in female gonads is at least a few hundred times higher than in male gonad during the period between 55–127 dpf at 8–10°C. Hence down-regulation of *aromatase* gene expression in the male gonad is a critical step in sex differentiation in the normal male and as well in the artificially sex reversed phenotypic female. Guiguen et al. (1999) have also shown that aromatase activity in the gonads of the normal female is much higher at 64 dpf fry than that of a masculinized XX neomale flounder *P.olivaceus* fry fed with aromatase inhibitor or MT and concomitant down-regulation of ovarian *aromatase* gene expression (Kitano et al., 2000).

2.6c *Danio rerio*

O. latipes and *O. niloticus* are male heterogametic and primary gonochores, in which gonadal differentiation proceeds from undifferentiated gonad directly to the ovary and testis. Based on gynogenic (but not cytogenic) evidence adduced by Uchida et al. (2002), the zebrafish *D. rerio* is regarded as female heterogametic, *albeit* doubts still remain on its gamety. Braat et al. (1999) have shown that the migration of PGCs is completed and cluster with precursor PGC supporting somatic cells are recognizable at 5 dpf. The PGCs proliferate between 5 and 10 dpf to produce about 30 PGCs; however, Braat et al. (1999) remain silent on any difference in the number with regard to the possible presumptive sex. In *G. aculeatus*, a stickleback characterized by differentiated gonad, proliferation of PGCs precedes directing the ovarian differentiation (Lewis et al., 2008). In zebrafish embryo, meiosis occurs as early as 10 dpf and oocytes appear between 14 and 29 dpf (Takahashi, 1977; Maack and Segner, 2003). Gonadal differentiation in the male involves a transformation from juvenile ovary to testis. Briefly, the ovarian development is initiated by 10 dph and progresses until 20 dph. Between 15 and 25 dph testis development is initiated in presumptive males, simultaneously with apoptosis of ovarian cells including the ovarian cavity (Jorgensen et al., 2008). For details on the mechanism of apoptosis, Krumschnabel and Podrabsky (2009) may be consulted. It may also be indicated that many authors have endeavored to search for the gene or genes (Bradley et al., 2011) responsible for sex differentiation in zebrafish with a differentiated gonad. This book has made an attempt to trace the sequence of mile stone events in the search for sex determining/differentiation in zebrafish.

In the absence of any established biochemical marker for genetic sex of zebrafish, Uchida et al. (2002) have used a morphological feature, the percentage of perinuclear oocytes (see Fig. 30, Pandian, 2011) as an index to distinguish a presumptive male from a female. They have found that the percentage of perinuclear oocytes with regard to the total number of germ cells in an 'ovary' is significantly lower in the presumptive males than in the females. Accordingly, the percentage increases from 8% in a 15 dph fry to 17% in a 29 dph presumptive male fry, in comparison to the increase from about 20% to 44% in a female fry during the corresponding period. Hence Uchida et al. have claimed that sexually dimorphic differentiation has indeed commenced at 15 dph, when the difference in the percentage of perinuclear oocytes between the presumptive males and females has become apparent. Hence the occurrence of juvenile hermaphroditism is evidenced by the presence and persistence of oocytes and the ovarian cavity in presumptive males; it is however transient for a short period of a few days; yet their formation precedes the induction of male sex differentiation.

Using a *vasa:: egfp* transgenic line to identify sex transforming juveniles, Wang and Orban (2007) have shown that the expression of *cyp19a1a* commences in the non-transforming juvenile; it is up-regulated around oocytes in the ovary until the 3 wpf. During this period of 1–3 wpf, *cyp11b* and *amh* are not detectable. But in the gonad of the transforming juvenile, the expression of *cyp19a1a* is down-regulated from 3wpf onwards and the *amh* expression is increased by 300–600 times by up-regulation in regions, where *cyp19a1a* has been expressed earlier. Expression of *cyp11b* is also up-regulated but later than *amh* and its localization is not related to the position of degenerating oocytes. Briefly, at 3 wpf, i.e., at 21 dph dimorphic sex differentiation becomes more apparent and established, as evidenced by *egfp*-positive females and *egfp*-negative presumptive males. Apparently, *amh* is one of the prime candidate genes for down-regulation of *cyp191a* leading to the transformation of juvenile ovary into testis. Hence the peptide hormone *amh* inhibits the expression of *cyp19a1a* (cf Rodriquez-Mari et al., 2005) and is responsible for the degeneration of oocytes along with the ovarian cavity, a male specific event (Uchida et al., 2002). Notably, Guo et al. (2005) have shown the strong expression of *Dmrt1* in the perinuclear oocytes but a fainter one in matured oocytes. Remarkable differences in the number of developing oocytes in presumptive male zebrafish have also been recently reported (Hsiao and Tsai, 2003). These findings on the molecular genetic basis of male differentiation in zebrafish seem to confirm the earlier observations of Uchida et al. that the difference in percentage of perinuclear oocytes may serve as a morphological index of male differentiation.

Incidentally, Wang et al. (2007) have reported that presumptive males, in which the differentiation process towards the male commences at the 21 dph, approach completion of testicular differentiation around 32nd, 35th and 43rd dph (Fig. 15). Krovel and Olsen (2002, 2004) have found that the *vas:: egfp* transgene is strongly expressed in gonads of the female but only faintly in that of the male zebrafish. Wang et al. have subjected 500 or so individual zebrafish at 28 dph for an intensive investigation. They have found wide variations among the tested individuals not only at the commencing point and duration of *egfp* expression but also in its intensity

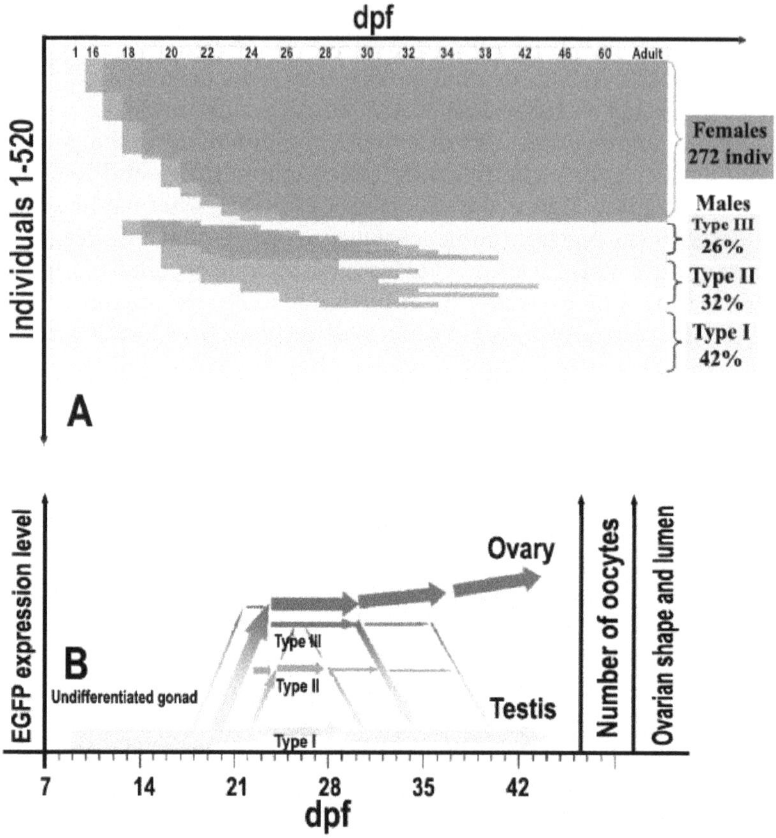

Fig. 15. *Danio rerio:* (A) Upper panel: The zygotic *vas::egfp* commences to highly express after 16 dpf in females but males of types I, II and III, which began to faintly express, progressively decrease the expression at different dpf (day post-fertilization). The increasing fluorescence intensity suggests the increasing oocyte number and growth of ovarian lumen and shape. In contrast the decreasing fluorescence indicates the degeneration of oocytes and transformation of ovary into testis. (B) Lower panel: A model for gonad differentiation showing high levels of variation in the intensity, onset and duration of gonad transformation (from Wang et al 2007. The timing and extent of the juvenile ovary phase are highly variable during zebrafish testis differentiation. J Fish Biol, 70A: 33–44, reproduced by permission of The Fisheries Society of the British Isles/John Wiley & Sons Ltd)

Color image of this figure appears in the color plate section at the end of the book.

(Fig. 15). Considering the intensity of expression, shape of gonad and number of oocytes, the transforming presumptive males are grouped into three types (Table 15). At least 42% of presumptive males seem to have completed male differentiation at the 32 dph.

Sex differentiation in juvenile/adult zebrafish is mediated by genetic signals from autosomal genes. Expression profiles reported by Jorgensen et

Table 15. Classification of gonads of presumptive male zebrafish at 28 dph (compiled from Wang et al., 2007)

Type	Features
I	EGFP expression too low to be detected. Neither the ovary-like cavity nor the elongate shape of ovary found before or after transformation. Oocytes are not detected or hardly with one or two in one sectioned ovary. Comprises 42% of presumptive males, which directly develop into males
II	Faint EGFP expression. Number of oocytes are few and are less than that observed in type III. In terms of structure of the ovarian cavity and shape of ovary, gonads are partially developed. Comprises 32% of presumptive males
III	Normal EGFP expression. Oocytes are detected. Comprises 26% of presumptive males

al. (2008) for six genes suggest an important role for *ar*, *sox9a* and *Dmrt1* in controlling male differentiation and *fig a* and *cyp19a1a* in regulation of female differentiation (Fig. 16). Evidence from studies on the non conventional fish pejerrey *Odontesthes bonariensis*, in which sex differentiation is known

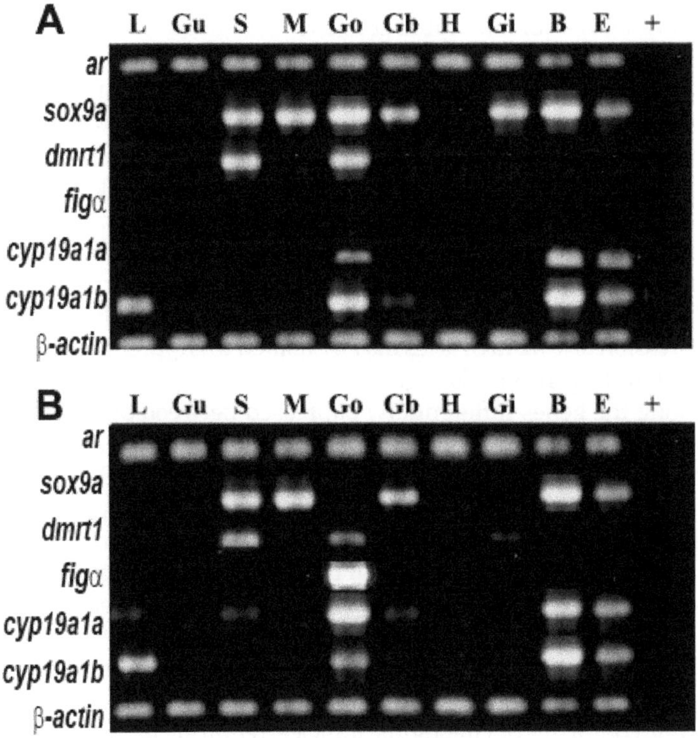

Fig. 16. Tissue specific expression of *ar, sox9a, dmrt1, figa, cyp19ala* and *cyp19alb* from (A) male and (B) female zebrafish. L: liver, Gu: gut, S: spleen, M:muscle, Go: gonad, Gb: gall bladder, H: heart, Gi: gills, B: brain, E: eye, +: control (from Jorgensen et al., 2008)

to depend on temperature (Strussmann et al., 1998), as in zebrafish, has shown that anti-apoptosis gene *pen-2* plays a specific role in protecting germ cells from apoptosis (Fernandino et al., 2010; see also Silva et al., 2008). Knockdown of *pen-2* strongly induces the p53-dependant apoptosis cascade (Cambell et al., 2006). Further Fernandino et al. (2010) have also observed a higher expression level of *amh* at masculinizing temperatures than at feminizing temperatures during the gonadal differentiation. At intermediate temperatures, the expression level of *amh* remains higher in half of the larvae than in the others. *Cyp19a1* shows a reciprocal expression pattern.

The zebrafish has been used as a model species to assess the effects of endocrine and similar chemicals on reproduction. Yet the molecular mechanisms controlling zebrafish reproduction are poorly understood. Using a 17K oligo nucleotide microarray, Santos et al. (2007) have investigated the molecular basis for the phenotypic specificity between sexes, between individuals within a sex and reproductive status of breeding zebrafish. The gonadal transcriptome has been found to differ substantially between sexes. Microarray analysis of individual gonad reveals the consistent expression of about 8,769 genes in the gonads, from which 7,976 and 7,060 are consistently expressed in ovaries and testes, respectively. Of them, statistical analysis has identified 2,940 genes that are differently expressed between males and females; among these sex specific genes, 1,570 are overexpressed in females and 1,370 in males. Gene ontological analysis has revealed that *sox11b, sox21a* and *sox31* are overexpressed in ovaries; in contrast, the absence of elevated expression of *sox9a* and the relative overexpression of *amh* and its receptor in testes induce not only the differentiation and development of testes but also the maintenance of the sexual dimorphism. Incidentally, the overexpression of *foxl2* in ovaries may be related to suppression of biological activity of androgens in this organ.

Among the genes differentially expressed between ovaries and testes in the array dataset, 53 genes have been further identified to show a 10-fold difference in expression between sexes. A magnitude of 3,000-fold difference in expression of some genes between the ovary and testis is likely to play a crucial role in context of sex specific gonadal function. Variation between gonadal transcriptomes of individual males is limited and may account for the phenotypic difference in sperm quality, such as sperm motility. However, the variation between individual females exceeds the levels observed for the males. Within the growing follicles, two groups are identified; the first one includes predominantly early oocytes undergoing oogenesis and the second one comprises of vitelloginic follicles, requiring batteries of enzymes (e.g., cathepsins) to process this large yolk precursor into small molecules, which are stored within the oocytes to provide nutritional support for the developing embryos. Cluster analysis of 13 genes selected from these females with predominance of vitelloginic follicles are perhaps involved in arresting prophase I of the meiotic division of most germ cells in the early oocyte stage. This may be a reason for the reductions in relative fecundity and relative fecundity rate observed in aged female fishes (cf. Fig. 27).

In recent years a number of reports on the mutants in zebrafish reversing its sex have been published. For instance, Abrams and Mullins (2009) considered only the maternal defective mutants, indicating the occurrence of >40 unique mutants to confirm the maternal control of early differentiation in zebrafish. Table 16 lists some of these mutants relevant to sex reversal during embryonic and post-hatching stages in zebrafish.

Table 16. Selected examples of mutants and their role in sex differentiation in zebrafish

Mutant	Remarks
Ziwi	Reduces the number of PGCs (cf Saito et al.,2008) leading to production of all male progenies (Houwing et al., 2007)
Fancl	Expresses in developing germ cells of bipotential gonads at the critical time of sex differentiation. Caspase 3 immuno assay reveals the increased apoptosis of germ cells in homozygous mutants leading to production of all male progenies. On failure of oocytes to survive through meiosis, the PGC supporting somatic cells fail to express the ovarian gene *cyp19la* nor the consequent down-regulation of the testis gene *amh*. However, the introduction of tumor suppressor gene *tp53* into *fancl* mutants rescues production of females, indicating that *fancl* is involved only with survival of developing oocytes through meiosis (Rodrequez-Mari et al., 2010)
$Brca^{Q658X}$	Homologue of *BRCA2* mutant in human with hereditary breast and ovarian cancer. $Brca^{Q658X}$ induces the failure of embryonic differentiation of oocyte of zebrafish. Oogenesis proceeds normally but produces binucleate oocytes (Shive et al., 2010)

2.6d *Silurus meridionalis*

Only fragmentary information is available on genes responsible for morphological and molecular sexual dimorphic differentiation in this catfish. Yet, it seems to provide an interesting but a different differentiation process and calls for further study on the role of sex differentiation genes from zero dah. Information on the gamety of the catfish, as well as whether the catfish belongs to the tertiary gonochore, in which potential intersexual gonad, develops directly into either an ovary or a testis is required. In gonochoric teleosts, the initial ovarian ontogeny varies with species and is first indicated either by the appearance of the ovarian lumen or by germ cell meiosis (Nakumara et al., 1998). In *Cichlasoma dimerus, O. latipes,* and *A. anguilla,* the entrance of oogonia into meiotic division and development of primary growth stage oocytes occur prior to the formation of the ovarian lumen but it is after the formation of the lumen in *P. olivaceus, P. lethostigma, O. niloticus* and *O. aureus* (see Luckenback et al., 2003). Apparently,

S. meridionalis belongs to the second group. Incidentally, the oogonial entry into meiosis and formation of the ovarian lumen occur simultaneously in *Carassius auratus, Cottus bairdii* and *O. mossambicus* (Nakumara et al., 1998).

In an attempt to trace the cause for all-female production in the Southern catfish *S. meridionalis*, Liu et al. (2008) have described a histologically sex-wise indistinguishable gonad until 8 dph at 24°C. Remarkably, the female gonad commences two processes, which gradually fuse to form an ovarian cavity at 12 dph fry, a clear sign of morphological sex differentiation. In the male gonad, the cavity is not visible at 12 dph. Hence *S. meridionalis* may not belong to secondary gonochore like the zebrafish, in which the ovarian cavity is formed along with oocytes in all fry known as 'juvenile hermaphrodites'.

Notably, the proliferation of PGCs commences only in the 29 dph fry, clearly after the ovarian cavity is formed on the 12 dph; the PGCs enter into the fast proliferation phase in 35 dph fry and end with the terminal number of 712 in a 85 dph fry (Fig. 13). In the male gonad, they also commence around the same time and reach the fast phase in a 85 dph fry ending with the terminal number of 1,159 PGCs (Zhang, 2005). Hence there is a clear difference between the chronological sequences of dimorphic sexual differentiation in medaka and tilapia on one hand and catfish on the other: The number of PGCs is significantly higher in the male than in female catfish, indicating perhaps the relatively lower expression of the *Dmrt1b* in the male catfish; this is in contrast with the presence of larger numbers of PGCs in female medaka and tilapia.

Liu Z et al. (2007, 2008) have also estimateed the levels of *cyp 19, foxl 2* and *Dmrt1a* and *b* in the developing gonads, unfortunately at a later period of differentiation. High levels of expression of *cyp19, foxl2* and *Dmrt1a* and *Dmrt1b* in the female gonad at 65 dph, when the sex dependent PGC proliferation has just become apparent but certainly later, after the formation of the ovarian cavity at 12 dph. However, the expression levels of *Dmrt1a* and *Dmrt1b* are higher testes. Moreover, the expression of *cyp19b* in the brain pituitary and gonad is suggestive of its role in B-P-G axis (Lui Z et al., 2007). Besides *gthα, fshβ* and *lhβ*, known for their expression in zebrafish (So et al., 2005) and protandrous gilthead sea bream *Sparus aurata* (Wong and Zohar, 2004), are expressed in the catfish, especially the *gth α* and *fsh β* from 25 dph and *lhβ* 40 dph again later than the day, when the ovarian cavity is formed.

2.7 Genetic cascades

Figure 17 is a schematic representation of the genetic cascades through which the dimorphic sex differentiation is realized in the selected models

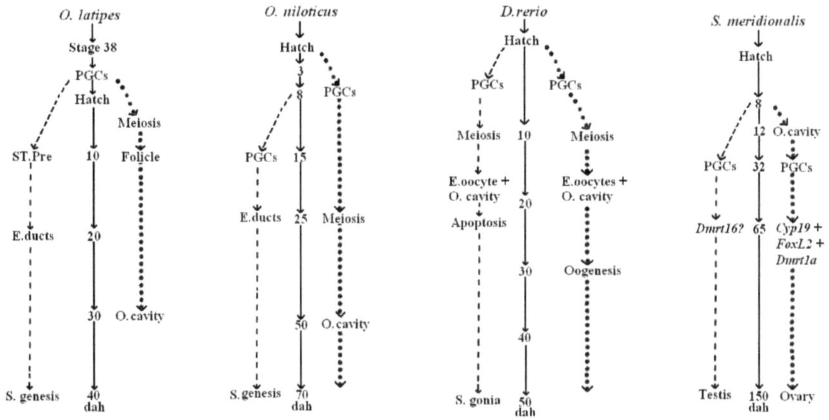

Fig. 17. Genetic cascades of sex differentiation in the selected model fishes. ST Pre = Sertolli cell pre-cursors, E.ducts = Efferent ducts, S. genesis = Spermatogenesis, O. cavity = Ovarian cavity, E. oocytes = Early oocytes. For the Southern catfish *Silurus meridionalis* alone, genes, known to be involved in sexual differentiation are shown, as they are not considered elsewhere

of *O. latipes, O. niloticus, D. rerio* and *S. meridionalis*. The differentiation becomes apparent prior to hatching in *O. latipes* but in the others a few days after hatching. In females, sexual differentiation proceeds from meiosis in germ cells through the formation of follicles and the ovarian cavity, and to oogenesis much earlier than the respective males. As indicated, early oocytes and ovarian cavity are formed in all the juvenile hermaphroditic zebrafish, but between the period from 15 and 29 dph, the oocytes and cavity are degenerated by apoptosis in presumptive males; subsequently testicular differentiation is completed. Whereas *O. latipes* and *O. niloticus* are primary gonochores, the zebrafish is a secondary gonochore. It is not yet known whether the Southern catfish is a tertiary gonochore, in which the juveniles initially possess a bipotential gonad, which then develops directly into either an ovary or testis, as in *G. aculeatus* (Lewis et al., 2008) and *Anguilla anguilla* (Grandi et al., 2010). The chronological events in the genetic cascades are bound to vary among these primary and secondary gonochores. In comparison to gamety, not much is known about which are the species that belong to the primary and secondary gonochores. However, even within a genus, for instance, *Oncorhynchus*, one species is a primary gonochore (e.g., *O. kisutch*) but the others like *O. masou* are secondary gonochores (see Pandian, 2011). Hence tracing the chronological sequence of genetic cascade even among biologically and economically important fish may prove to be an onerous task.

Following Morrish and Sinclair (2002), Ijiri et al. (2008), Siegfried (2010) has made an admirable attempt to sequence the timing of gene expressions in gonads of *O. latipes, O niloticus* and *D. rerio* (Fig. 18). The timing and

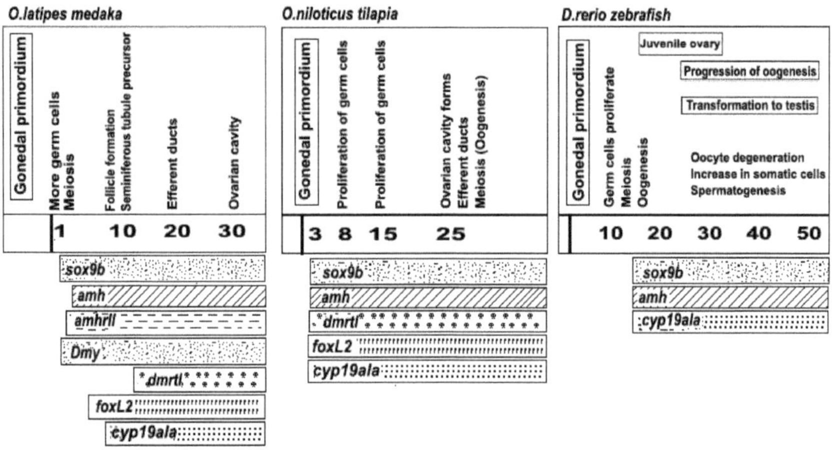

Fig. 18. Comparative illustration of timing of gene expression in gonads of *Oryzias latipes,Oreochromis niloticus* and *Danio rerio* during gonadal sex differentiation. The white bar represents a time line of developmental age (dph/dpf). Hatching is indicated by a vertical slim bar in the white bar. Land mark events in the gonadal differentiation are indicated above the white bar within boxes. The timing and duration of expression of each gene are indicated below the bar in horizontal rows (from Siegried 2010. In search of determinants: genes expression during gonadal sex differentiation. J Fish Biol, 76: 1879–1902, redrawn and reproduced by permission of The Fisheries Society of the British Isles/John Wiley & Sons Ltd)

the sequences and their sexual dimorphic expressions vary considerably among these species. However, there are many similarities between gonadal gene expression in these three species and in comparison to mammals they reveal the conserved role of many genes during vertebrate gonadal sex differentiation. The mutations inducing sex reversal at the PGC proliferation and the results of experimental hormonal induction of sex reversal described here clearly indicate that changes at the earlier stages of the genetic cascades are tolerated but not at the later stages. Notably the genes at the top of the cascade are not conserved, whereas the downstream genes have been homologues in a much broader spectrum of species (Herpin and Schartl, 2011). The scope for wide diversity and complexity in timing and strength of expression of one or the other genes in the cascade renders enormous plasticity and flexibility in dimorphic sexual differentiation in fishes.

2.8 Gonadal ducts

The reproductive system consists of two functional units: gonads and genital ducts. The gonads generate gametes and the genital ducts serve to transport the gametes to the exterior. In mammals and elasmobranchs, the genital ducts develop in close relation to the renal duct systems. The paranephric or Mullerian duct differentiates into oviduct, uterus, cervix and upper vagina and the mesonephric or Wolffian duct differentiate into vas deferens, epididymis and seminal vesicles (van Tienhoven, 1968). In fish, the ovarian duct consists of (i) the intra-gonadal duct, which is continuous with the oviduct and (ii) the extra gonadal. In the male the tube-type testis contains an efferent duct, which extends from testis and functions as the extra-gonadal duct (Suzuki and Shibata, 2004).

Considering *O. latipes* as a model for fish, the development of genital ducts, as described by Suzuki and Shibata (2004) is briefly summarized: in both sexes, the extra-gonadal genital ducts contain two structural units: the anterior and posterior parts. The anterior part of oviduct extends continuously from the ovarian cavity and ovary (Fig. 19). The posterior

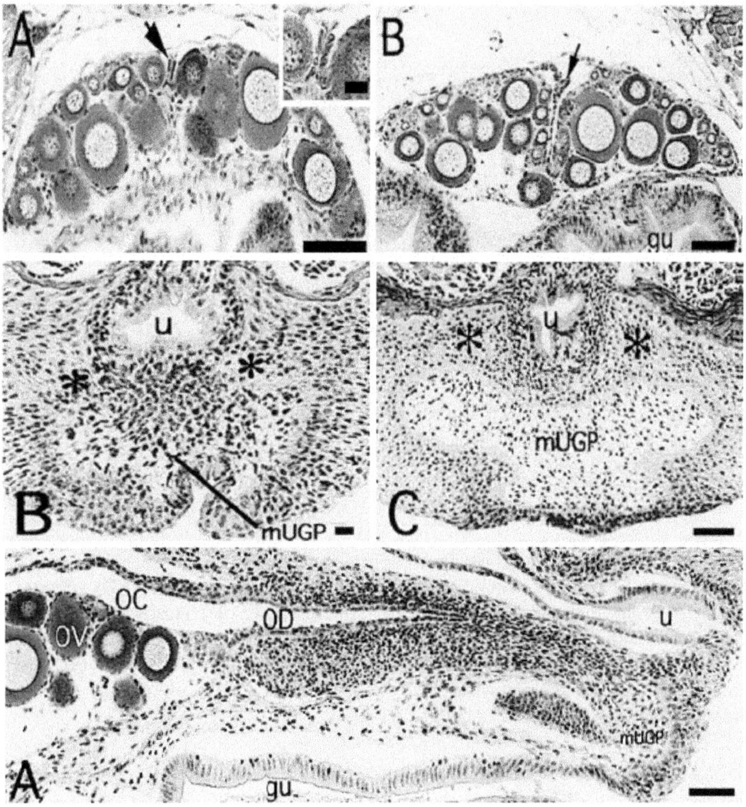

Fig. 19. Development of gonadal ducts in *Oryzias latipes*. (A) Transverse section (TS) of posterior end ovary, arrow shows cell mass of dorso-central part of ovary; (B) the same but the arrow showing the presumed primordial ovarian cavity; (C) Elongation of genital pore lip (GPL): TS of UGP at body lengths of 15 mm and (D) 20 mm; (E) Vertical section of mid sagital plane, OV = ovary, OC = Ovarian cavity, OD = Oviduct, U = urethra, gu = gut, mUGP = medulla of urinogenital papilla, scale = 50 μm (from Suzuki and Shibata. 2004. Zoological Science, 21: 397–406)

part of oviduct, namely the genital pore lip (GPL) results from invagination and caviation of the cortex of uninogenital papillae (UGP) and forms the oviduct opening (Fig. 20).

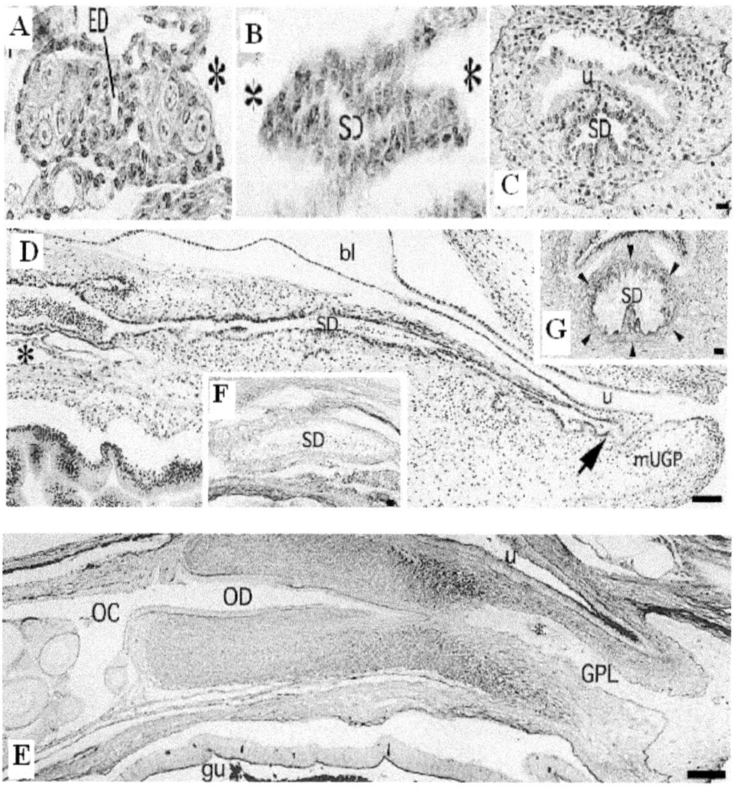

Fig. 20. Development of efferent duct and sperm duct in *Oryzias latipes*. TS of testis of a fish with body length of 12 mm showing the appearance of (A) efferent duct (ED), (B) sperm duct (SD) and (C) development of urethra (U) scale = 10 μm, (D) TS of the testis to urinogenital pore in a fish with 25 mm body length. mUGP = medulla of urinogenital papillae bl = bladder (E) For comparison, formation of genital pore bp. (GPL) cavities in female with body length of 24 mm is shown. OC = Ovarian cavity, OD = oviduct, gut = gut scale = 50 μm (from Suzuki and Shibata. 2004. Zoological Science, 21: 397–406)

Both male and female gonads are bilobed at 10 to 20 dph. Formation of genital ducts is initiated in both sexes at 20 to 30 dph. Development of UGP and extra-gonadal ducts is related; the UGP consists of the corex and medulla. The ventral region of urethra mesenchyme may have an important role in the formation of the extra genital duct (Fig. 19). Land mark events in anatomical and histological sex differentiation are briefly summarized in Table 17. Onikate (1972) reported that administration of estrone induced early formation of anterior oviduct in female medaka. But treatment with

Table 17. Land mark events in anatomical sex differentiation in ovary and testis in *Oryzias latipes* (compiled from Suzuki and Shibata, 2004)

Age (dah)	Ovary	Testis
20–30	Cell layers at posterior end of ovary are extended. Cell layers and mass, ventral to ovary differentiate into ovarian cavity and oviduct (Fig. 19A, B, 20E)	PGC supporting somatic cells develop as acinus precursors, which differentiate into efferent duct and sperm duct (Fig. 20B). Urethra is developed ventral to sperm duct (Fig. 20C).
31–50	UGP of medulla becomes apparent and is connected with mesenchyme cell layer between anus and urinary pore (Fig. 19C,D)	Typical UGP medulla is observed. Posterior end of urethra (Fig. 20D).
51–90	Two layers originate from each side of cortex. They invaginate into ventral region of the mesenchyme and form cavities namely genital lip, a female specific feature (Fig. 19C,D). Fusion of the cavities open into oviduct anteriorly and to exterior posteriorly. Prominent UGP is bi-or trilobed (Fig. 19E) GPL is located dorsal to UGP medulla.	Spermatogenesis and sperm are observed. Epithelia of anterior sperm duct and urethra are nearer to urinary pore (Fig. 20D).

fadrozole, an aromatase suppressant, severely inhibited the formation of ovarian cavity and anterior oviduct (Suzuki et al., 2004).

Incidentally, *amh/mis* is responsible for the regression of Mullerian ducts in the mammalian male fetus. The role of *mis* in gonadal sex differentiation in fishes, which have no Mullerian ducts, remain unclear. Shiraishi et al. (2008) have shown that *mis* and *mis* type2 receptor (*misr2*) mRNA are expressed in somatic cells surrounding the germs cells of both sexes during sex differentiation of medaka. Germ cell proliferation is suppressed by the loss of either *mis* or *misr2*. When tissue fragments containing germ cells are treated with recombinemt eel *mis*, the germ cells proliferate more significantly in both sexes, compared with the untreated control. In contrast, culture tissue fragments from the *mis* or *misr2* defective embryos inhibit the proliferation of germ cells in both sexes. In medaka, *mis* indirectly stimulates germ cell proliferation through *misr2* expressed in somatic cells immediately after they reach the gonadal primordium. However, *Dmrt* may play an important role in dimorphic sexual differentiation of germ cell proliferation.

Though inappropriate, differentiation of genital ducts in a protandric is briefed to provide a continuity. The black porgy *Acanthopagrus schlegeli* has a striking life cycle, including an early sex differentiation, bisexual gonad and male to female natural sex change. According to the description by Lee et al. (2011), the gonads develop 'ovarian cavity cracks' and become 'four stranded' at the age of 16 weeks. Subsequently, genital ducts are drawn as an extension of the ovarian cavity, and are extended to form the oviducts. The testicular cavities are also extended and fused together to form the outer duct namely the sperm duct. The origin and course of events in the formation of the gonadal ducts in this protogynic resemble those in the gonochore like *O. latipes*. During the natural sex change, the male reproductive passage of the porgy is regressed and degenerated. The oviduct, arrested as a blunt and during the male phase, is finally extended and connected to the genital pore during the female phase (see also Alam and Nakamura, 2007).

2.9 Puberty and maturation

Puberty is the transitional period between the immature juvenile stage and reproductive adult stage, at which an individual begins to produce the first wave of rapid spermatogonial proliferation or the first batch of oocytes accumulating cortical alveoli. The two main functions of gonads are the production of fertile gametes and synthesis of reproductive steroid hormones, implying functional competence of brain-pituitary-gonad (BPG) axis. These are convincing evidences that the regulatory input from outside the gonad is required to trigger pubertal development. Only in captively reared fish, the progressive changes in gonadal histology (e.g., Holland et al., 2000; Bernet et al., 2009) and endocrine profile (e.g., Haukenes et al., 2008) can be followed. Expectedly, this has implications for farmed fishes. Taranger et al. (2010) have reviewed the problems and prospects of puberty in farmed European fishes. In most fishes, it is initiated earlier in males (Table 18), for instance, *Acipenser sinensis* male matures at the age of 14 years, whereas the females by 21 years. Interestingly, Insulin like Growth Factor (IGF1) is a key somatotropic hormone that conveys growth status to the reproductive endocrine system. For instance, IGF1 increases steady-state mRNA levels of *gnrhr1* and four other gonadotropin sub-units *fshb, chb, cga1* and *cga2*, and may act alone or with GnRH to increase pitiuitary FSH release in coho salmon *O. kisutsch* one year before puberty, i.e the IGF1 may prime pituitary gonotropic cells of pre-pubertal salmon to respond to GnRH by stimulating the synthesis of GnRH and FSH at the onset of puberty (Luckenbach et al., 2010). The onset of puberty is marked by the release of pituitary-gonadotropin on stimulation by GnRH. Kiss proteins, the peptide products of the *kiss-1* gene and its receptor (Gpr54) are the gate keepers of GnRH functions. Extensive studies undertaken on the puberty of the European sea bass have suggested that *Gnrh1* and its receptor *dlGnrh-II1a* are most relevant for the onset of puberty (Taranger et al., 2010).

Oviparous fishes can be separated into two groups according to their spawning strategy. The semelparous species have a single spawning event during their life time, e.g., salmon (Crespi and Teo, 2002) and a few

Table 18. Selected examples of age/size at maturation and spawning patterns in teleosts

Species/Reference	Age/Size at maturity	Remarks
Oryzias latipes, seasonal spawner (see Lam, 1983) *O. javonicus*, annual spawner (see Lam, 1983)		Spawns every day
Danio rerio, Balasubramani and Pandian (2008a)	120 ± 5 days	Spawns once a week
Betta splendens, Balasubramani and Pandian (2008b)	140 days	Spawns every fortnight
Morone saxatilis Holland et al. (2000)	♂ 1–3 years ♀ 3–7 years	Puberty initiated in 65% ♂ Puberty initiated in 65% ♀
Acipenser schrencki, Wei (1997) Bernstein et al. (1997)	♂ 7–8 years ♀ 9–10 years	Spawns once every 4 years
A. medirostris Erickson and Webb (2007)	♂ 153–188 cm ♀ 167–192 cm	
A. sinensis Wei et al. (1997), Deng et al. (1991)	♂ 14 years ♀ 21 years	86% ♂ milted at 14+ years 76% ♂ spawned at 21 years; season/multiple spawner
Argyrozona argyrozona Brouwer and Griffiths (2005)	♂ 30–40 cm ♀ 30–40 cm	Spawning frequency decreases from 9 to 4 times, as size increases
Nothobranchius rachovi (see Pandian, 2010)	6 week	Spawns once in life time
Coregonus lavaretus Bernet et al. (2009)	♂ 78–981 dph 15–19 cm ♀ 678–981 dph 19–21 cm	

cyprinodontids (Pandian, 2010). The iteroparous species employ one or the other of the following strategies: (i) the annual single spawners, i.e., species which spawn repeatedly during their life time: e.g., *D. rerio*, and (ii) the annual multiple spawners, i.e., species which have a long life span and spawn once or several times during the breeding season of each year of their life time (Nunez and Duponchelle, 2009).

The onset of sexual maturity represents a critical transition in the life of an individual, as allocation of time and resources has to be provided for reproduction, besides survival and growth. Indeed maturation marks metamorphic changes in 'endocrine climate'. As age and size at maturity can have significant consequences, it is important to know whether the age or body size determines sexual maturation. Using adequate data on growth, fecundity and life span of populations of 19 fish species, Stearns and Crandall (1984) developed a model to predict whether age or size determines sexual maturity in fish. The model accounts for 82% of the variants in the age/size of maturity. Accordingly, when growth is either rapid or slow, the fish may have a fixed size, at which they mature. Conversely, when growth is intermediate, the fish may have a fixed age, at which they mature. There are also other predictive models, which account for 60% variations in the age of first reproduction for 31 populations representing 23 species (Roff, 1992). A large body of literature is available on variations in age and size at which, 50% of the collected sample fishes are identified to have attained sexual maturity. Table 18 lists such variations reported for a few selected examples of fishes, which are known for their different modes of spawning.

Causes for such wide variations in age/size at maturity are related to social and/or environmental factors. Age at puberty can be delayed or advanced, depending on the environmental conditions; for instance, both age and size at puberty have been reduced in several heavily exploited populations (e.g. Engelhard and Heino, 2004, Olsen et al., 2005; see also Pandian, 2010), indicating the phenotypic plasticity or genetic changes due to size selective fishing and thereby reducing both age and size at puberty (Dieckmann and Heino, 2007). A reason for sexually dimorphic growth and differences between size and age at puberty (Table 18) has been traced to different relationships between body size and reproductive success in males and females (Taranger et al., 2010). Selection often favors early maturing individuals, because their higher survival to maturity and relatively greater contribution to population growth. But it may also favor delayed maturation, when fitness increases with age/size at maturity (Schreibman and Kallman, 1978). A sex linked gene called *P-locus* is known to primarily control the age/size at maturity. On maturity, body growth of poeciliid males is almost arrested; consequently, poeciliid males are usually 60–80% the length of the female and are more slender (Bisazza, 1997). Therefore, the age/size at maturity of poeciliids has attracted a lot attention to assess the role of the social environment on *P-locus*. Kolluru and Reznick (1996) designed two long term experiments to assess the inheritance of age and size of males at maturity. In the poeciliid, *Phallichthys quadripunctatus* males attain sexual maturity either at a small size of 15 mm body length or 19 mm size. Briefly, large males sired early and delayed maturing sons; likewise, the smaller males too sired delayed and early maturing sons. Hence the size at maturity in this poeciliid is more effectively controlled by social environmental factors than *P. locus*.

Temperature is a very important factor in determining growth and maturation. For instance, the European cod *Gadus morhua* matures at the age of 3+ in the relatively warmer British waters but in colder Barents Sea within Arctic Circle, it begins to mature at the age of seven years (see Wooton, 1998). In this context, a recent publication by Bernet et al. (2009) is interesting. In a commendable description, they provide the chronological sequence of events in gonad differentiation upto maturation of females and males as functions of age and size with due consideration to temperature in the whitefish *Coregonus lavaretus*. From Fig. 21, selected events in sexual differentiation are

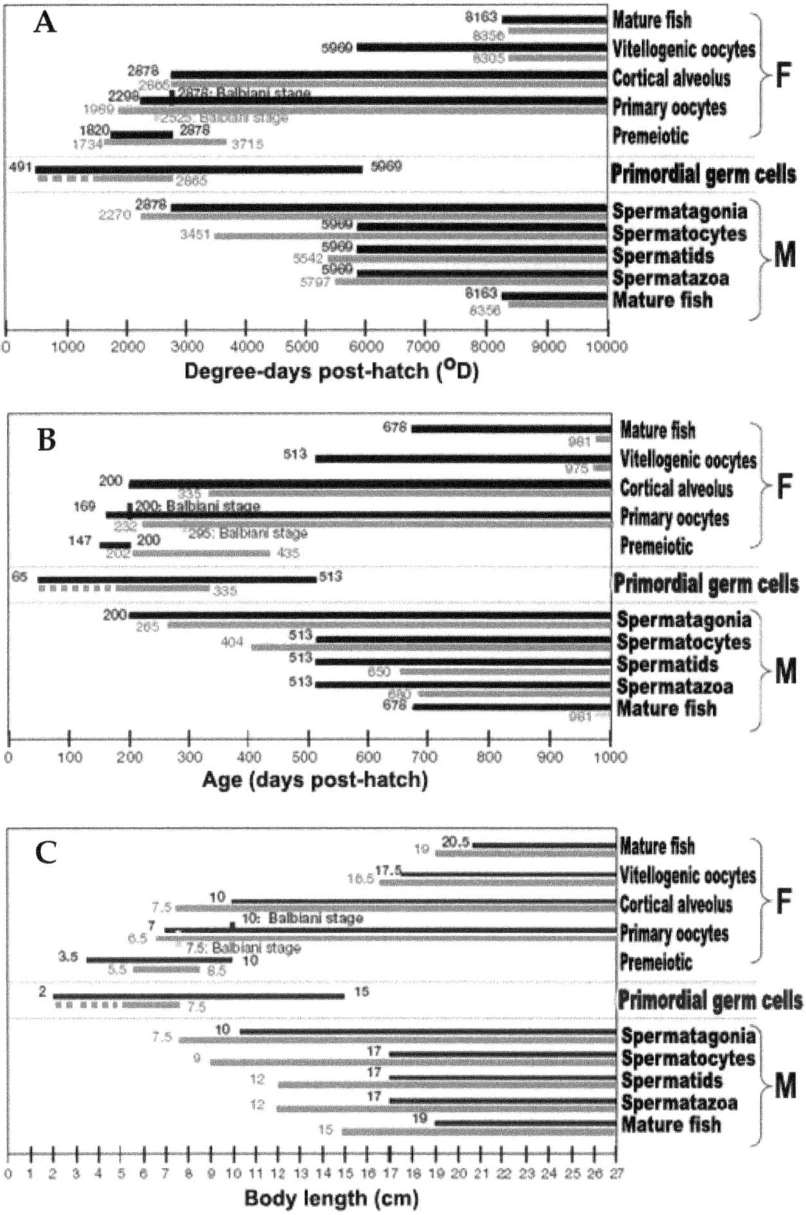

Fig. 21. *Coregonus lavaretus* sex differentiation from the arrival of PGCs to maturation as functions of (A) degree-days post-hatch, (B) age dph, (C) body length of fry at Lake Thun (temperature 4–22°C, black bars) and spring water (8–9°C, gray bars). Numbers near the bars represent the first (left) and/or last (right) occurrence of a cell type stage (from Bernet et al., 2009)

highlighted in Table 19. From a long term well designed experiment, Bernet et al. have undertaken at two sites, one at the natural waters of Lake Thun, Switzerland and the second one at constant temperature of 8–9°C at a spring water area. They have traced the ovarian differentiation as interrupted by a shorter or longer period of resting phase, while the testicular differentiation once commenced proceeds fast and continuous without any interrupting resting phase. However, the thermal requirement of about 8,200°D is almost equal for maturation of male and female, irrespective whether the fish have been reared at constant (8–9°C) temperature, as in spring waters, or at habitat thermal regime of 4–22°C as in the Lake Thun. Nevertheless, the slow rate of differentiation in spring waters lasts till maturation; consequently, the difference in the events of age of sexual differentiation becomes wider and wider, as they grow. Hence the difference in age at sexual maturity between the Lake Thun and spring water groups is 303 days or 45% variation. In terms of size at sexual maturity, this variation amounts to 23% for males but only 8% for females. Therefore, temperature affects the age at sexual maturity more effectively than size at sexual maturity; size at male maturity is more affected than that of the female.

Table 19. Chronological sequence of events leading to ovarian and testicular differentiation and sexual maturation in *Corgonus lavaretus* (compiled from Bernet et al., 2009)

Event	At Lake Thun 4 to 22°C			At Spring water 8–9°C		
	Age (dph)	Size (cm)	Degree -days (°D)	Age (dph)	Size (cm)	Degree -days (°D)
Undifferentiated gonad and arrival of PGCs	65	2.0	491	65	2.0	491
			Ovarian differentiation			
Oocyte development	147	3.5	1820	202	5.5	1734
Perinuclear oocyte	169	7.0	2298	232	6.5	1989
Balbaini body stage	200	7.5	2878	295	7.5	2878
Resting phase	313			640		
Vitelligenic oocyte	513	17.5	8356	975	16.5	8305
Matured	678	20.5	8163	981	19.0	8356
			Testicular differentiation			
Spermatogonia	200	10.0	2878	265	7.5	2270
Spermatocytes	513	17.0	5969	404	9.0	5542
Resting phase		No			No	
Spematozoa	513	17.0	5969	404	9.0	5542
Matured with milt	678	19.0	8163	981	15.0	8356

2.10 Acrosomeless sperm and micropylar eggs

Among vertebrates, teleost fishes are unique for the lack of acrosome in their spermatozoa (Jamieson, 1991). Despite much modification in shape and mitochondria to suit oviparity and viviparity, the spermatozoa do not have acrosome (Fig. 22). This feature of lack of the acrosome is unique and common to teleost fishes, *albeit* the rudimentary acrosome is reported in the sturgeon *A. baeri* (Psenicka et al., 2010). Hence, the gene(s) responsible for the lack of acrosome is conserved among the fishes. However, no research is as yet available on this feature.

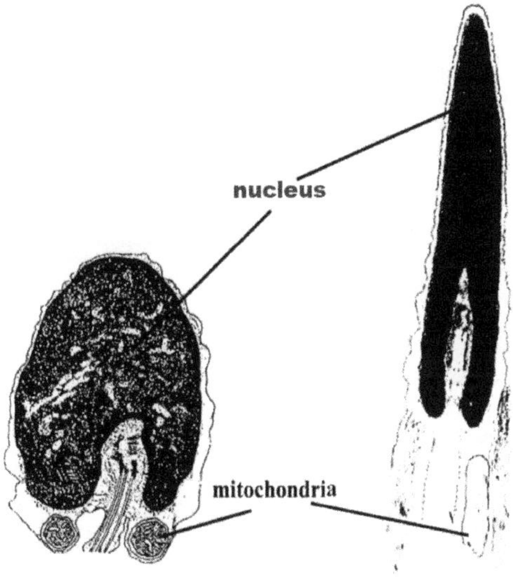

Fig. 22. Longitudinal sections of aquasperm of oviparous *Oncorhynchus tshawytscha* (left) and interosperm of viviparous *Poecilia latipinna* (right). Note the mitochondria in *O. tshawytscha* sperm and the large modified ones in *P. latipinna* (from Jamieson, BGM. 1991. *Fish Evolution and Systematic Evidence from Spermatozoa*, Cambridge University Press).

Similarly, a single micropyle is present in the egg of fishes. The sperm enters through micropyle of the egg (see Pandian, 2010, Fig. 23). This has facilitated heterospecific fertilization in many fishes (Pandian and Kirankumar, 2003) and also led to hybridization and polyploidization (Pandian and Koteeswaran, 1998), which may modify the process of sex determination and consequently sex differentiation. Hybridization is known to occur between more than 300 species (Argue and Dunham, 1999). Hence a detailed account on structure and function of micropyle is provided. Incidentally, the presence of micropyle in eggs is unique and common to teleost fishes; research on identification and characterization of genes responsible for the formation of micropyle in fish eggs is also desired.

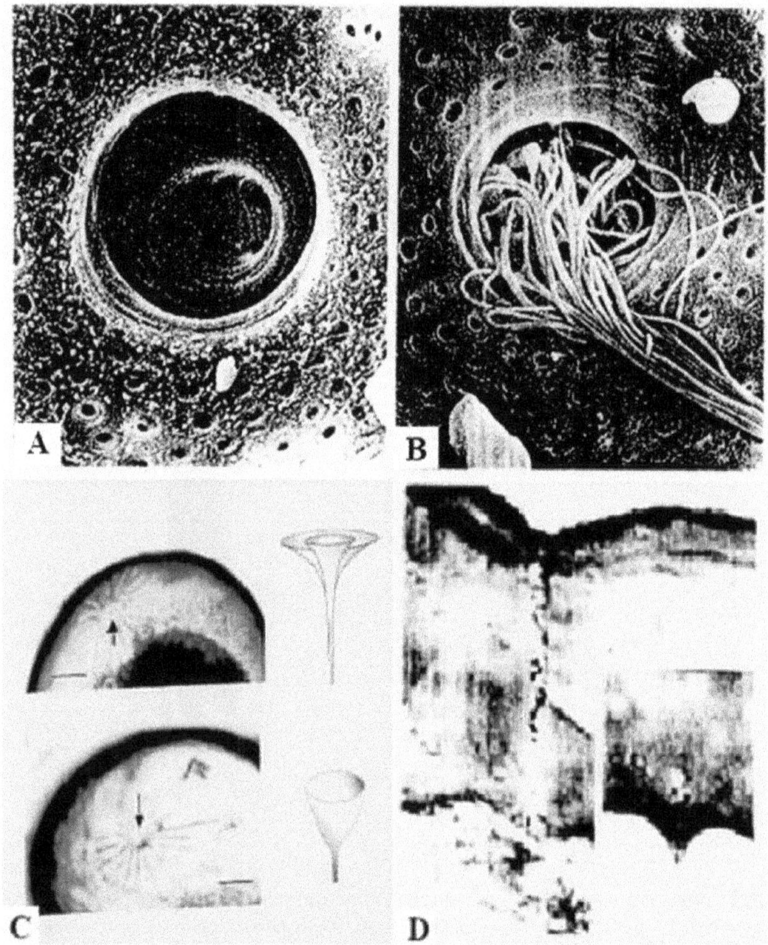

Fig. 23. Electron microscopic views of (A) egg showing micropyle and (B, D) entry of sperm through the micropylar canal in an egg of *Alcichthys alcicornis* (from Munehara et al., 1989). Apical view of unfertilized eggs of (C) *Gymnocorymbus ternetzi* (lower) and *Hemigrammus caudovittatus* (upper) and schematic views of their respective micropylar canals in eggs (from David, 2004)

Ravaglia and Maggese (2002) have described the structure of micropyle in eggs of the swamp eel *Synbranchus marmoratus*. The ovulated single oocyte, drawn from the ovarian cavity, has a compact vitelline envelope surrounding the oocyte with a single micropyle as the only entrance in the animal pole (Fig. 24A). The vitelline envelope in the vegetative pole has a polyhedrical surface that resembles 'floor tiles' (Fig. 24B). Over the equatorial region, these tiles change their shape and elongate to form isolated crests. Near the animal pole, these crests fuse together and become

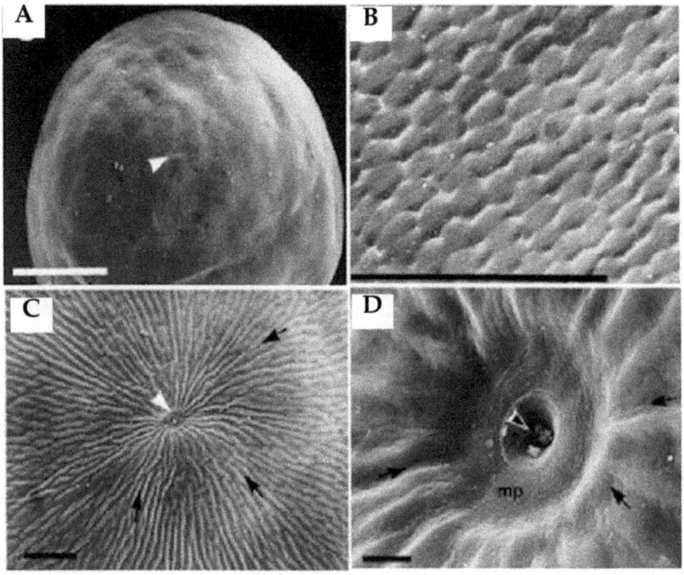

Fig. 24. *Synbranchus marmoratus*: SEM pictures of ovulated oocyte: (A) Apical view, (B) Vegetative pole surface view, (C) Magnified view of animal pole showing furrow-like structures that converge (indicated by arrows) in the micropyle and (D) Micropylar canal opens in the center of the micropyle pit. Scale A to C = 100 µm and d = 10 µm (from Ravaglia and Maggese, 2002)

furrow-like structures with a slightly spiraled direction (Fig. 24C). These furrows converge directly into micropyle pit. The micropylar canal is located in the center of this pit (Fig. 24D) (Ravagla and Maggese 2002).

Information on morphological patterns of furrow architecture guiding sperm to the micropylar pit is available for the eggs of many fishes. For instance, an incredible chorion architecture has been described with an array of ridges and furrows in the animal pole, all of them with spiraled arrangement and convergence into the micropyle canal of *Luciocephalus* sp. (Riehl and Kokoscha, 1993). In the catfish *Sturiostoma aureum,* Riehl and Spatzner (1991) have found eggs, whose vitelline envelope has furrows running from vegetal to the animal pole. Using time-lapse video and image analysis of sperm movement in the eggs of cyprinid *Barbus conchonius,* Amanze and Iyengar (1990) have estimated that the grooves guide 99.7% sperm into the micropyle penetration and/or fertilization.

The length and diameter of the micropylar canal differ from species to species (e.g., Blanc et al., 1993). Yet successful heterospecific inseminations occur, for instance, between *Cyprinus carpio* and *Ctenopharyngodon idella,* suggesting the canal's diameter and length are similar in both of them. Such reciprocal insemination is prevented in other species. For instance, the long canal with narrow anterior opening in the eggs of Bueno Aires tetra

Hemigrammus caudovittatus prevents the entry of sperm of the widow tetra *Gymnocorymbus ternetzi* (Fig. 23C). Conversely, the short canal with broad anterior opening in the eggs of *G. ternetzi* readily permits the sperm entry of *H. caudovitatus* (David and Pandian, 2006).

Interestingly, the chondrostean gametes differ basically from those of teleost fish in having an acrosome in spermatozoa (Psenicka et al., 2007) and multiple micropyles in eggs (Cherr and Clark, 1985), for instance, 10 in the hagfish *Eptatretus stouti* (Morisawa and Cherr, 2002). Hence the function of the acrosome, which undergoes acrosomal reaction and filament formation remains unclear, although the acrosome seems to increase the fertilization process (Psenicka et al., 2011). The mature egg of the paddlefish *Polyodon spathula* has four to 12 micropyles in the animal polar region (Fig. 25). The sperm entry site in the egg surface under the micropyle consists of tuft of microvilli. A minute after fertilization, a ball-like enlarged full-grown fertilization cone is formed, closing the opening of the micropyle (Fig. 25) and thereby preventing entry of more sperm (Linhart and Kudo, 1997). It is not clear whether the multiple micropyles with smooth outer surface of the chondrostean egg are replaced by a single micropyle in teleostean egg decorated with furrows and/or tile-like architecture to guide the sperm to the micropyle. A comparative study of the genes responsible for the development of a single apical micropyle in the egg and the loss of acrosome in the sperm may prove to be rewarding.

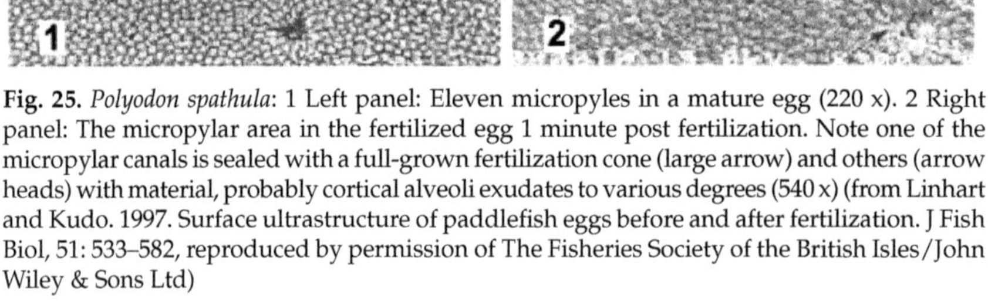

Fig. 25. *Polyodon spathula*: 1 Left panel: Eleven micropyles in a mature egg (220 x). 2 Right panel: The micropylar area in the fertilized egg 1 minute post fertilization. Note one of the micropylar canals is sealed with a full-grown fertilization cone (large arrow) and others (arrow heads) with material, probably cortical alveoli exudates to various degrees (540 x) (from Linhart and Kudo. 1997. Surface ultrastructure of paddlefish eggs before and after fertilization. J Fish Biol, 51: 533–582, reproduced by permission of The Fisheries Society of the British Isles/John Wiley & Sons Ltd)

2.11 Secondary sex characters

In many fishes, males and females are phenotypically very distinct and these differences often reflect divergent selective pressures acting on the sexes. Phenotypic sexual dimorphism reflects differing patterns of gene expression between the sexes. For instance, Small et al. (2009) have identified as many as 5,899 genes to express differently in males and females of *D. rerio,* despite the apparent absence of heterogamety. Table 20 lists selected examples of sex specific external secondary sexual characters. Hundreds of similar examples may be recognized from a survey of any standard book on 'Aquarium Fish'. These externally distinguishing characters are non-invasive biomarkers, very much sought to quickly recognize sex in a given commercially and biologically important fish. In sturgeons, the external sexual dimorphism is less obvious. Vecsei et al. (2003) have found the shape of external urinogenital opening of a live male is like the letter Y, while it looks like 'O' in a female. Using this non-invasive criterion, sex of 82% of live specimens can precisely be determined but not in dead specimens. Another non-invasive technique employs Enzyme Immune Assay (EIA) (i) to identify the mature female from the level of female-specific plasma vitellogenin (Vtg) and (ii) to distinguish the immature female from the male using plasma 17β estradiol and 11-KT levels in specimens of the giant air-breathing Amazon fish *Arapaima gigus.* Adults are 100% accurately sexed

Table 20. Selected examples of secondary sexual characteristics of fish

Characteristics	Remarks
Permanent	
Sexual size dimorphism	
Males > females	Macroandry, polygamous cichlids
Females > males	Microandry, poeciliids
Color	Male attractive, Female attractive, Red and yellow male morphs, purple-bellied female, *Pelvicachromis pulcher*
Jaw	Male specific elongation in salmonids
Dorsal bend	Female specific in poeciliids
Barbel	Larger barbels in male e.g. *Loricaria parva*
Pelvic fin	Fused to form anal pouch in female *Corydoras aeneus*
Dorsal fin	Male specific *Betta splendens*
Caudal fin	Male specific *Betta splendens* Male specific elongated caudal sword, *Xiphophorus helleri*
Anal fin	Gonopodium in poeciliids
Gonadal papillae	Male specific in gobiids, *Symphodus ocellatus*
Gestation pouch	Male specific in syngnathids
Light organ system	Male specific in leiognathids, *Photoplagios rivulus*
Transient	
Ovipositor	*Rhodeus sericeus, Pseudoblennius percoides*
Color	*R. sericeus,* male: olive emerald green on the back, orange-red on throat and belly, bright red on dorsal and anal fins, and green on caudal fin; female: yellowish green

by Vtg EIA, while 100% of immature females and 95% of adult males are accurately sexed by 17β estradiol and 11-KT ratios (Chu-Koo et al., 2009).

With reference to macroandry, the identification of *growth hormone pseudogene* in male salmonids is relevant (Table 9) but more details on this aspect are provided elsewhere (Pandian et al., 2012). Interestingly, on transplantation of Spermatogonial Stem Cells (SSCs) into the sterile alevins of triploid masu, the fertile xenogenic masu also displays nuptial body coloration and males with an elongated jaw, both of which are typical secondary sexual characters. For the other sex specific characters, more details can be obtained from Pandian (2010). The genetic mechanism of differentiation of most of these secondary sexual characters need to be studied. Nevertheless the morphological and molecular differentiation of gonopodium has received attention since 1938. Viviparity is usually facilitated by an intromittent organ namely gonopodium in Poeciliidae, andropodium in Herniramphidae, priparium in Phallostethidae, penes in Cottidae, tubular genital papillae in Athernidae or anal fins in *Corynopoma riisei* (Burns et al., 1995). The anal fin of goodeids is modified into a primitive gonopodium, which is not structurally and functionally similar to that in poeciliids (Bisazza, 1997).

Secondary sexual characters generally appear at puberty. The formation of gonopodium in poeciliids offers an entry point to understand the sexual differentiation in juveniles. Anal fins of male and female juveniles (<10 mm) of *Gambusia affinis* are identical (Turner, 1941). When a male reaches a body length of 15 mm and larger size, the anal fin rays 3, 4 and 5 are greatly elongated with increased number of segments from about 15–20 to 38–43; besides, the 4th and 5th rays are bifurcated, and adorned with elbow, serrae, spines and terminal hooks (Fig. 26), while all the other rays remain short. In the female, the fins remain short but the 4th and 5th rays are bifurcated and a little elongated. More details on cell proliferation, and further differentiation are described by Ogino et al. (2004).

In *G. affinis,* Ogino et al. (2004) have also cloned two *ar α cDNA* and identified them as *ar α* and *ar β*. Both are strongly expressed prominently in the distal region of outgrowing anal fin rays; *ar α* is expressed in both epithelium and mesenchymal regions of the anal fin, while that of *ar β* in the distal mesenchyme and outgrowing fin rays. The signaling molecule, *sonic hedgehog* (*shh*) induces the development epithelium of distal anal fin rays. *shh* participates in differentiation or regeneration of fins with *Ptc1* receptor, whose expression is also detected adjacent to the *shh*-expressing region. Administration of androgen induced anal fin outgrowth concomitant with the *shh* induction to the formation of a gonopodium. The functions of *ar* in the formation of intromittent organ in other fishes are indicated by Ogino et al. (2011).

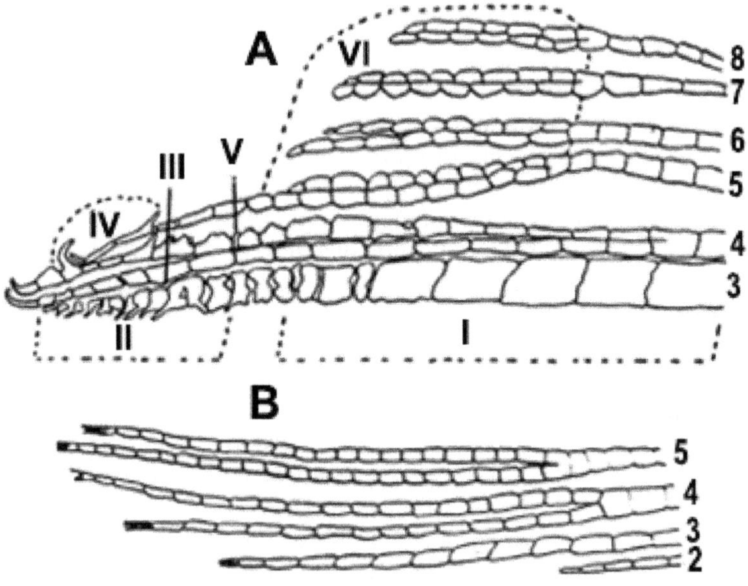

Fig. 26. Morphological differentiation of anal fin rays of *Gambusia affinis* Arabic numerals indicate fin rays number from the ventral side of the fin. Roman numerals indicate the location of the differentiation areas. Note the differentiation in fin rays number 3,4,5 in a (A) male and the same in a (B) female (from Turner, CL. 1941. Biol Bull, 80: 371–383, reprinted with kind permission from the Marine Biological Laboratory, Woods Hole, MA)

Color and pigmentation play an important role in camouflage, warning, theromoregulation, protection from ultra-violet radiation and courtship display. For instance, 99% males of *G. holbrooki* are silver and the remaining 1% is melanic. Melanic males have greater survival and larger gonopodium (Horth et al., 2010). Tezuka et al. (2011) have reported that among the melanics of *G. reticulata*, genetic variations in *melanocortin 1 receptor* (*MC1R*) gene coutribute to polymorphism of melanin pigmentation. Determining the complete coding sequence of *MC1R*, two alleles are recognized in *G. reticulata*, the one with 963-bp is less melanic than the other with 969-bp. However, the relation between *MC1R* genotype and black pigmentation disappears in the F_2 progenies.

Sabbah et al. (2010) have brought indicative evidence for the neural basis for sexually dimorphic visual behavior, i.e., females and males sample their visual environment differently, suggesting that the difference in detection of optical signals has a role in sexual selection in tuning color vision. A large number of publications are available on this topic, especially with reference to the explosive speciation rate among the African cichlids (e.g., Allender et al., 2003). Male body coloration is an important discriminatory factor used by females in mate choice. To gain insight into the molecular basis of cichlid coloration, Gunter et al. (2011) have compared transcription in the bright blue skin of males to yellow skin in females of the Malawi cichlid *Pseudotropheus saulosi*. Among the cDNA microarray-identified 46 clones that displayed expression differences between the two sexes, five have been confirmed to be differentially expressed by relative quantitative Real-Time PCR (qRT-PCR). This gene list includes a representative from the endosomol-to-Golgi vesicle trafficking pathway *Coatmer protein complex subunit zeta-1* (*Copz-1*), a critical determinant of pigmentation. Transcriptional differences and putative functional diversification of genes seem to cause the coloration of blue males and yellow females among the cichlids. Incidentally, van der Sluijs et al. (2010) have found that the coloration of progenies of *Pundamilia pundamilia* in the turbid areas of Lake Victoria is determined by the dames but not the sires.

Apart from these African examples, there are others describing how mate choice may lead to increased isolation between strains. For instance, Fukamachi et al. (2009) identified a medaka mutant (*colur interfere: ci*) that is distinctly unattractive to breeding partners. The *ci*, a mutation on the *somatolatin α* (*SLα*) gene, seems to reduce orange pigment color cells (xanthophores) in the skin. Though there are extremely strong mating preferences, the mutant and SLα-transgenic strains of medaka prefer to mate with partners from their strain leading to increased isolation between these strains.

2.12 Morphotypes and differentiation

Among vertebrates, the occurrence of an array of morphotypes within a sex is astounding, although this area of sexual differentiation is hardly covered by molecular biologists. The intense sperm competition has driven the males to develop an alternate strategy, in which males have been forced 'to make best use of a bad situation' (Taborsky, 2001). Some of these male morphotypes require explanation/definition. Accordingly, a territorial male constructs a nest or defends a territory and courts with the female(s) in his territory. Sneakers are relatively small males, which dart and steal spawning with territorial females, e.g., bluegill sunfish. Facing intense competition, the spermatozoa of sneakers swim faster than those of pirates, whose sperms are the slowest swimmers (e.g., *Telmatochromis vittatus*, Fitzpatrick et al., 2007). Satellites are slightly larger males and may be tolerated by territorial males; with the same body color and behavior, they may deceive the parental/territorials, as in some species like the bluegill sunfish. The satellites are not homogenous and the term has been indiscriminately used (e.g., Oliveria et al., 2001b,c, Table 21). For instance, among yellow morphs of the cichlid *Pelvicachromis pulcher*, a pair-forming male, who has lost his territory or partner, is also called a 'satellite' (Martin and Taborsky, 1997).

Table 21. Ontogenetic growth and transformation from sneaker to satellite/helper and to male status in *Symphodus* spp, *Julidochromis, Telmatochromis,* and *Leptoscarus* (source as quoted in Pandian, 2010, updated)

	Sneaker ♂	Satellite ♂	Territory ♂
		Symphodus ocellatus	
Size (cm) Body colour Urinary papillae	3.5–6.0 Brown White	6.1–8.0 Greenish White	8.1–9.5 Red-Blue White
		S. tinca	
Size (cm)	8.0–12.0	12.1–20.0	20.1–27.0
Colour	Dull	Dull	Bright
		S. roissali	
Size (cm)		4.5–8.0	8.1 -14.0
Colour		Dull	Bright
		Julidochromis marlieri	
Size (cm)		4.0–4.4*	4.8–6.7
		Telmatochromis vittatus	
Size (cm)	2.3–3.4	4.6–5.5	5.1–6.2+
		Leptoscarus vaigiensis	
Size (cm)		6–8**	8 - 18***
		Parablennius parvicornis ☺	
Size (cm)	7.9–8.3	8.4–10.0	10.2–12.1
		Salaria parva ☺ ☺	
Size (cm)	4.5–7.3	4.9–7.3***	7.4–11.5
Head crest index (0–9)	0.10	0.14	0.35
Genital papillae (0–9)	1.14	1.70	1.41

*helper, + Pirate ♂ 6.4–6.7 cm, **IP, ***TP; ☺from Oliveira et al. (2001a); Note: Sneakers are taken as satellites; there are also floaters measuring 10.1–12.0 cm size, ☺ from Goncalves et al. (2008) ☺ ☺ from Miranda et al. (2003) *** transitionals

Katano (1983, 1992) describes cannibalistic male and 'female' satellites, who intensely compete against a pair-spawning *Zucco temminscki*. Sub and helper males are reported among cichlids. Kuwamura (1997) has recorded that a haremic male cichlid holds in his harem a sub-adult helper belonging to another species. With the exception of the bluegill sunfish and the like, all these morphotypes namely helper, sub-male, sneaker, satellite and pirate within the male sex represent only ontogenetic stages of the territorial male (Table 21). But the sneakers and satellites precociously mature. Knapp (2004) indicates the differential levels of reproductive hormones as the cause for the appearance of different morphotypes. Oliveira (2006) lists some 16 species, in which the bourgeois, a common name for territorials and nest-holders, have titre levels of both 11-KT and T higher than those of parasitic (sneakers, satellites and floaters) males. In many social fishes, subordinate helpers assist breeders in progeny care. A subordinated helper male in a co-operative social organization of *Neolamprologus pulcher* shows reduced levels of 11-KT and T, compared to a submissive helper male (Bender et al., 2006). With excretion levels of high T and low 11-KT, the helpers pursue a 'stand-by strategy for breeding' to react quickly, if an occasion for breeding arises (Bender et al., 2008). Hence hormones are mainly responsible for the presence of these ontogenetic precocious/submissive male morphs.

Conversely, the other morphotypes such as parental and cuckolder, hooknose and jack, haremic and pairing male morphotypes have a genetic basis for differentiation. In the bluegill sunfish, parentals mature at the age of 7+ and use a territorial strategy. Despite receiving and sharing equal maternal genomes, the progenies sired by cuckolders precociously grow fast, mature earlier and die before they ever reach the size of the mature parental, i.e., there is no evidence that the cockolders ever become parentals themselves (Gross, 1982, 1984). The nest-tending parentals provide less care of their 'suspected' progenies; they may partially cannibalize or abandon their 'suspected' brood (Neff, 2003a,b). These observations indicate that there is a genetic basis for male differentiation into parentals and cuckolders and apparently, cockolders sire cuckolders only. Contrastingly, there is experimental evidence among salmonids that hooknoses do sire both hooknoses and jacks bearing XY genotype. Heath et al. (2002) provided the much desired first evidence from the sires of the crosses: neofemale (XX) x female (XX) and male (XY) x female (XX) of hooknoses of the chinook salmon *O. tshawystscha*. They reported that the XY hooknose sired 31% jack and 69% hooknose progenies, in comparision to 10% jack and 90% hooknose progenies sired by XX neomale hooknose; they attributed the presence of 10% jack to autosomal effects. This is an important area, where experimental evidences are urgently required to know (i) whether the salmonid jacks also sire jacks and hooknoses or jacks only, and (ii) whether the parentals of sunfish sire parentals alone or both parental and cuckolder progenies. The findings from these experiments shall pave the way to study and understand the different elements, constituting the sex chromosome in these salmonids and centrarchids, and the role of the elements in differentiation of these male morphotypes.

The presence of two morphs namely territorial (type I) and sneaker (type II) are described in *Porichthys notatus* (Bass, 1996). The sneakers are about 50% smaller but have nine-fold larger testes; they neither build nests nor guard eggs, they do not produce the 'hum' calls like the territorials but produce grunts like females. Development of their vocal motor circuitry as function of age in these two morphs has shown that the sexual differentiation in these morphs have a genetic base (Brantley and Bass, 1994). Similarly, there are also larger and smaller morphs in *X. maculatus* (Schreibman and Kallman, 1977) and *X. nigrensis* (Ryan and Causey, 1989). Sexual maturity in them is controlled by a single gene; their size, age at sexual maturity and morph are genetically determined.

A more complicated but interesting morphotype is found in the Nigerian cichlid *P. pulcher* (Martin and Taborksy, 1997). The cichlid is small, sexually dimorphic with color polymorphism. The sex ratio ranges from 0.33♂ : 0.67♀ in the upper wet region of the Sombreiro River to 0.5♂ : 0.5♀ in the lower dry region. Females are intensely colored with a purple belly. In males, there are yellow and red colored morphs. These colors are genetically determined. Red morphs are more aggressive and about 50% of them are haremic and hold two or more females in each of their harems. The other 50% red morphs are pair forming males. Individual red males can change between haremic and pair-forming status but cannot change into yellow morphs. The yellow morphs develop into pair-forming males, or become satellites, when the partner or territory is lost. The yellow morphs cannot change into red ones, although they can change between paired and satellite status. Apparently, the body color of these male morphs are sex-linked. However, it is not known which gene(s) in the Y chromosome induces the differentiation of the body color and a host of associated behavioral traits (see Pandian, 2011). From the point of genetic sex determination and differentiation, it is important to know that all the observed morphotypes among gonochores are phenotypic variants and fall within a sex, say male or female, i.e., they are all intrasexual morphotypes only, as these morphotypes do not cross the borders of a sex. Contrastingly, such phenotypic variants among hermaphrodites are intersexual, as they do cross the borders of sex.

2.13 Reproductive senescence

As mentioned elsewhere, sex determination refers to that specific event lasting for seconds and minutes only, when a heterogametic gamete fuses with that of an opposite sex resulting in the development of an ovary or testis in a bipotential gonad. But the process of sex differentiation is continued until death of the vertebrates. This consideration opens a larger issue of whether the fishes, being poikilotherms, undergo reproductive senescence or not. Fishes display extraordinary diverse patterns of ageing. For example, the Pacific salmon undergoes very rapid senescence at spawning, whereas rockfishes, characterized by indeterminate growth, exhibit very slow or negligible ageing (Patnaik et al., 1994). Many small fishes with a short life span such as zebrafish, medaka and guppy are characterized by regular time-dependant ageing similar to other vertebrates (Gerhard et al., 2002). Fishes do accumulate many age-related biomarkers like β-galactosidase, oxidized proteins and non-degradable lipofuscin (e.g., Ding et al., 2010), as in other vertebrates. Genetic up-regulation of cholinergic transmission attenuates cognitive impairment in zebrafish (Yu et al., 2006). Koshimizu et al. (2011) describe the possibility of inducing embryonic senescence and laminopathies in zebrafish harboring disturbed expressions of the *lamin A gene* (*LMNA*). Translational blocking of the *LMNA* induces apoptosis, cell cycle arrest and cranio-facial abnormalities. Hence the induction of embryonic senescence by cryptic splicing of *LMNA* reduces the life span from 35–40 months in *zprogerin* transgenic zebrafish to 25–35 months in *zpregerin* transgenic. Incidentally, studies on the ageing process covering the entire life span are required to have a better understanding of reproductive senescence. But "there have been few studies, which have set out specifically to investigate the ageing process throughout the life span of any fish" (Woodhead, 1978). "Little has changed since" (Gerhard, et al., 2002). Besides there are also wide differences in the life span among different strains within a species. For example, mean life span of the outbred zebrafish is 42 months with the longest longevity of 66 months. But the *golden sparse* zebrafish survives for 56 months with the longest longevity of 58 months. Resvertol, a natural phytoalexin present in grapes and red wine is shown to increase longevity of the short-lived fish *Notobranchius furzeri* from about 13 weeks to 18–22 weeks by retarding the expression of age-related markers of senescence (Valenzano et al., 2006).

The asymmetric mitotic cystic division is the source of oocytes and Oogonial Stem Cells (OSCs) (Fig. 32). Hence the rate of egg/litter production by a female fish may reveal whether the generation of oocytes and OSCs is slowed down or ceased, with advancing age or increasing body weight. This may also serve as a signal for the onset of reproductive senescence. A voluminous body of literature is available on Fecundity (F) of fishes. Many of them show that Batch Fecundity (BF) is related to the volume of body cavity available to accommodate the ripe ovaries. As fishes vary enormously in their body shape, body weight, rather than size and age, is better correlated with F and BF (see Pandian, 2010). Hundreds of publications describe the relation of body size with F or BF, but have not provided body weight of the test fish (e.g., Thibault and Schultz, 1978) or their data are limited to a fraction of the early stage of life span; for instance, the reported data on body size-fecundity relation in *Poeciliopsis lucida* and *Poecilia reticulata* are limited to 30–31% of the life span of these poeciliids (Thibault and Schultz, 1978). Rarely, very few authors have provided all or most of the desired data; analysis of their data confirms that the fish do indeed experience reproductive senescence.

To estimate the number of broodstock required to regularly supply the desired number of fry, Coward and Bromage (1999) have maintained a stock of brooders of *Tilapia zilli*, a substratum (repeat) spawning cichlid of the size ranging from 32 to 461 g for a period of 80 days. Thankfully, they have reported data on BF, inter-spawning interval, relative fecundity

(RF) and egg size. From their data, it is possible to calculate the rate of egg production (number of eggs per gram fish per day) as function of body weight. Figure 27 shows the established positive correlation between body weight-BF relation. In the natural but man-made lakes of Cote d'Ivoire, the fecundity of the Nile tilapia *O. niloticus* also increases from 149 to 2,797 eggs for fish weighing 36 g to 975 g (Duponchelle et al., 2000). In fact such a positive relation between fecundity and body weight has been observed even among sea-run and resident strains of *S. leucomaenis*, who produce larger and smaller eggs, respectively (Morita and Takashima, 1998). However, the body weight-RF relationship in *T. zilli* clearly shows that the RF holds a negative relation to body weight (Fig. 27). Even more revealing is the relation between body weight and egg production rate (RFR), which

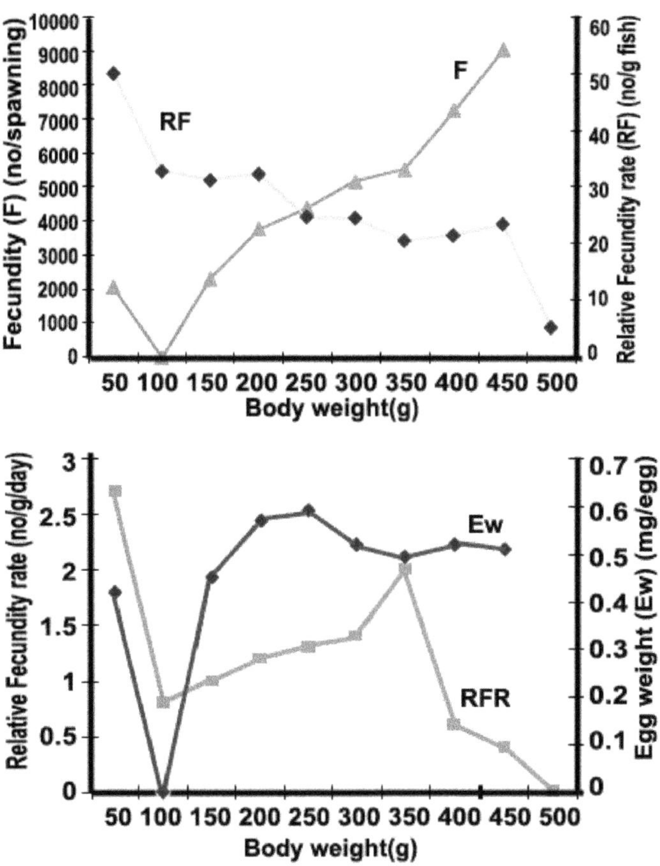

Fig. 27. Fecundity (F), relative fecundity (RF), relative fecundity rate (RFR) and egg weight (Ew) as function of body weight of substratum repeat spawner *Tilapia zilli* (compiled from Coward and Bromage, 1999)

is perhaps reported for the first time (Fig. 27). The trend shows that the egg production rate increases upto about 300 g body weight but beyond which it declines, clearly indicating that (i) the process of reproductive senescence is progressively expressed more strongly from 300 g to 461 g and (ii) the egg (and possibly oocytes + OSC) production rate also decreases from about 2.0 eggs/g/day in a 227 g fish to 0.1 egg/g/day in a female weighing 461 g. Of 13 larger females, nine (i.e., 69%) have not spawned even once during the 80 day period, though they are expected to spawn once every 56 days. Egg size is known to increase in larger females (see Pandian, 2010). A possible alternate is that the four larger females that spawned have produced larger eggs at the cost of reduced number of eggs, as is true of some marine fishes. In fact the follicle number per gram ovary is reported to decrease with increasing follicle (egg) size in many marine fishes including *G. morhua, Clupea harengus, Pleuronectes platessa, Scomber scombrus* and *Sebastes marinus* (Fig. 28). The egg size of *T. zilli* also shows a trend of increase but only upto 200 g size (Fig. 27). Clearly, *T. zilli* undergoes a terminal size related period of reproductive senescence.

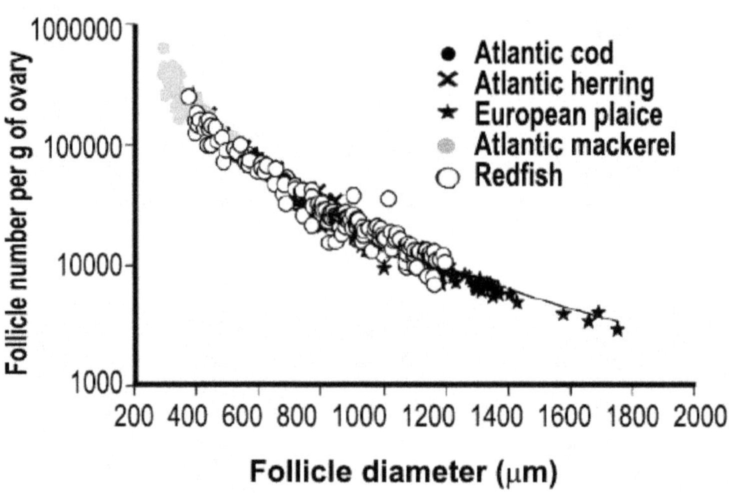

Fig. 28. Calibrated follicular diameter of some marine fish as function of relative fecundity (expressed in follicle number/g ovary) (from Witthames et al., 2009. Fish Bull, 107: 148–164)

The annual spawner *Hoplostethus atlanticus* is a long-living fish. It attains maturity at the age of 27.5 years (4,000 eggs/year), reaches the peak fecundity at age 125+ (16,000 eggs/year) and subsequently its fecundity is levelled around 6,000 eggs/year at age 150+; however, older females of upto the age of 175+ have been recorded (Minto and Nolan, 2006). The reason why Coward and Bromage (1999) have not found a significant correlation between age and fecundity is due to the presence of old senescent tilapia in their sample. Calculation of the data by Minto and Nolan shows that (1) fecundity holds a positive relation with body size (Fig. 29) but relative fecundity, which increases to the maximum at the age of about 125+, perceptibly decreases thereafter. The same holds true for the egg production rate RFR as well.

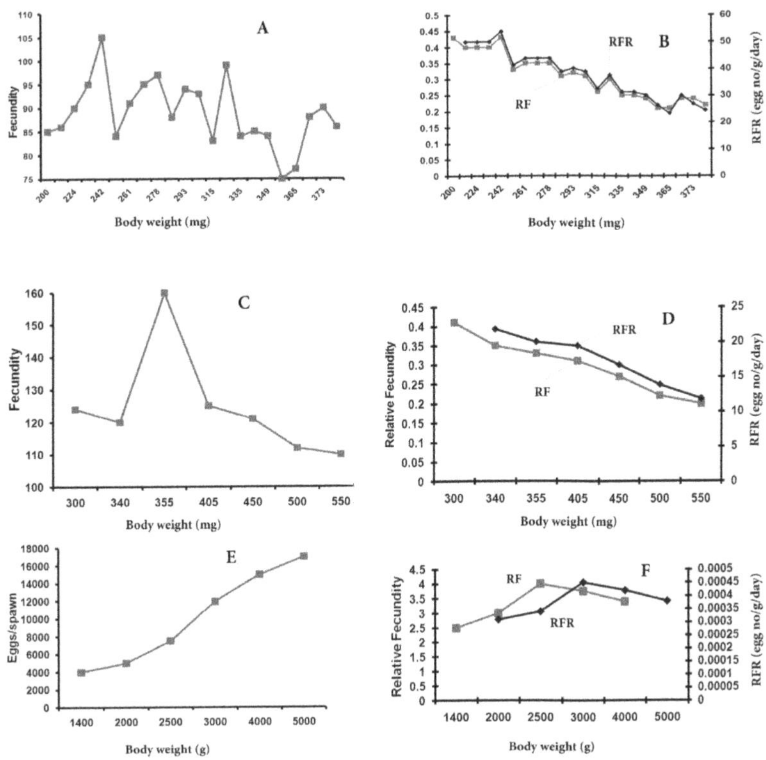

Fig. 29. Fecundity, relative fecundity and relative fecundity rate of small annual *Danio rerio* whose eggs are left uncared (A & B, compiled from Balasubramani and Pandian, 2009a), *Betta splendens*, whose eggs are guarded by male (C & D, compiled from Balasubramani and Pandian, 2009b) and large long living *Hoplostethus atlanticus*, whose eggs are left uncared (E & F, compiled from Minto and Nolan, 2006)

From a rare investigation undertaken on long living iteroparous rockfishes *Sebastes aleutianus* and *S. alutus*, de Bruin et al. (2004) recorded the continuation of oogenesis with follicular and oocyte enlargement in active ovaries of females even upto the age of 80 years. "Seventeen samples of mature (> 9 years) *S. aleutianus* ranging in age from 19 to 80 years and 248 samples of mature of *S. alutus* ranging in age from 9 to 77 years were obtained. Fortunately, *S. alutus* was sampled before spawning, when pre-ovulatory follicles are at maximum development" (de Bruin et al., 2004). Hence very rare and useful data have been made available for further analysis. Firstly, the distribution pattern of early follicles (200–350 µm diameter), small (351–500 µm) and larger (> 501 µm) oocyte size classes in > 50 year old *S. alutus* (the corresponding values for *S. aleutianus* are reported upto the age of 50 years only) shows that the early follicle size class was totally absent in females older than 50 years, implying slow or cessation of cystic division, and production of early follicles and incidental generation of stem cells after the age of 50 years. Correspondingly, the fecundity of *S. alutus,* which remained the highest around 300,000 eggs/year during the 40–50 years of age began to decline and reached the lowest fecundity of 180,000/year at the age of 77 years, suggesting the declining availability of early follicles to be recruited for vitellogenesis and further maturation. Unfortunately, the body weight data in relation to age of *S. alutus* are not provided to estimate relative fecundity and relative fecundity rate. However, the net annual increase in fecundity progressively decreased from 120,000 eggs at the age class of 20 years to 3,000–4,000 eggs in females aged between 70 and 77 years. Lastly, despite the vast data, the general inverted U trend suggests that the percentage of early follicles maturing to oocyte in *S. aleutianus* gradually declined from about 30% (number per 2 mm optical field) at the age of 20 years to about 17% at the age of 77 years (Fig. 30). These analyses of the data of de Bruin et al. (2004) allow the generalization that the rate of cystic division certainly begins to decline, if not totally ceases, with advancing age, resulting in the decreased production of oocytes and stem cells (see Fig. 32). The doctrine of confinement of stem cell genesis to pre-natal age in mammals is itself challenged (Johnson et al., 2004). Yet the doctrine of declining rates of cystic division and genesis of stem cells with advancing age shall remain unchallenged.

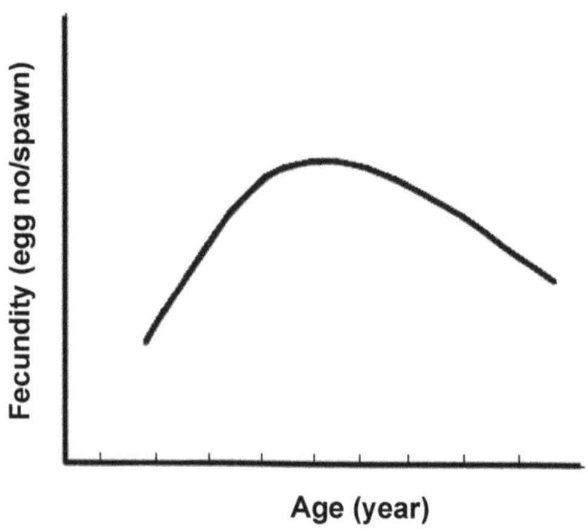

Fig. 30. Schematic representation of fecundity as function of age in long living fishes (compiled from de Bruin et al., 2004)

Similar trends confirming the onset of reproductive senescence in older females are also observed for the weight-F, weight-RF and age-egg production rate in *D. rerio* (Fig. 29), which do not afford parental care (Balasubramani and Pandian, 2008a), *Betta splendens* (Balasubramani and Pandian, 2008b; see Kishi et al., 2003), in which males guard eggs in the bubble nest until hatching (Fig. 29). Calculation of data reported by Thibault and Schultz (1978) suggests also a negative trend between body size and BF in *Poeciliopsis turneri* and *P. prolificia*, which are lecithotrophic and matrotrophic viviparous fishes. In their analysis of life history correlating to the evolution of viviparous fishes, Goodwin et al. (2002) noted that viviparous fishes produce larger offspring than oviparous fishes; despite their larger body size, the viviparous fishes produce offsprings only a tenth of that of egg layers. There are no other differences between them. Hence it is likely that the viviparous fishes also undergo a period of reproductive senescence.

Understandably the live-bearing guppy *P. reticulata* from a high predation environment mature at an early age and has a long reproductive life span (Reznick et al., 2006). Midlife cessation of ovulation, followed by a post reproductive life span of up to 33% of total life span has been documented for the guppy (Comfort et al., 1961). Guppies produce 15–28 litters once every 15–28 days. As they age, some females skip litters or even cease to reproduce for an interval of time, as has also been observed among the large females of *T. zilli* by Coward and Bromage (1999). This irregular cessation of reproduction accounts for the unusually long interbrood interval lasting for longer than 100 days in a few aged guppies. Briefly, guppies subjected to high predation spent a menopause period for 12% of their total life span, while those subjected to low predation have a longer period of menopause amounting to 15% of their life span (Fig. 31).

The described analysis of these rare publications not only confirm the occurrence of reproductive senescence and entrance of larger/older females into menopause but also suggest a minimal or cessation of OSCs production in old iteroparous fishes. It is not known whether they also occur in older migratory semelparous Pacific salmon, Atlantic eel and Indian hilsa and the non-migratory semelparous cyprinodontids. The semelparous tooth carps, which mature in 6 weeks *Nothobranchius rachovi* (see Pandian, 2010) and *N. furzeri* (Valdesalici and Cellerino, 2003) may serve as model to confirm whether they too experience reproductive senescence or not.

Incidentally, these findings on the onset of reproductive senescence in larger/older females have at least a couple of implications: (i) broodstock management in aquaculture and (ii) maintenance of OSC. Most interestingly, the capacity for egg production in relatively large fish like *T. zilli* (0.15–2.0 eggs/g/day, Fig. 27) and long-living *H. atlanticus* (0.00031–0.00038 egg/g/day, Fig. 29) is very limited. Contrastingly, the smaller fish like *D. rerio*

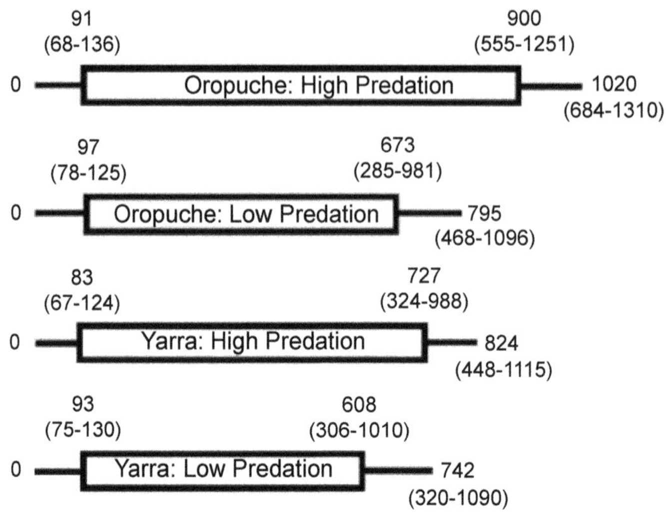

Fig. 31. *Poecilia reticulata*: Effects of high and low predation on age at maturity (front time lines in the left end), active reproductive life span (in boxes) and post-reproductive life span (time line on the right end). Arabic numerals indicate the mean values and ranges (in brackets). Note the post-reproductive life span is upto 824–1020 days and 742–792 days in populations subjected of the high and low predation, respectively (from Reznick et al., 2006)

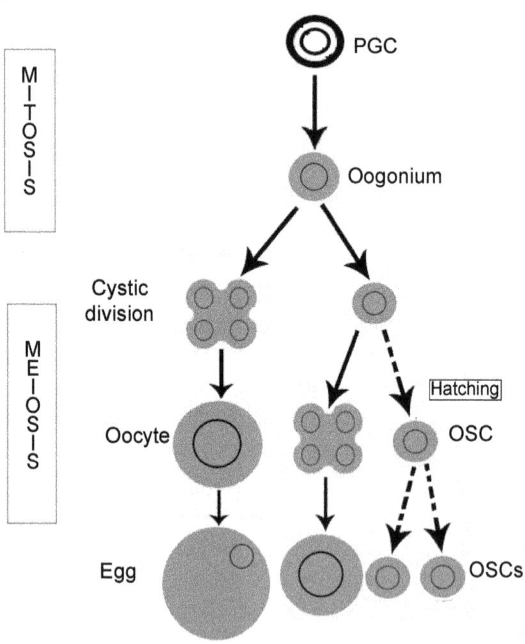

Fig. 32. Oogenesis in fish. Note the generation of germ cells during embryonic and adult stages

and *B. splendens* are capable of producing 24–50 eggs/g/day (Fig. 30) and 12–22 eggs/g/day (Fig. 30), respectively. The life time fecundity of *H. atlanticus* is about 0.15% of its terminal body weight, and that for the experimental period of *T. zilli* is 0.30% of its final body weight. In contrast, similar estimates for *B. splendens* and *D. rerio* are in the range of 1.5–2.6% of their respective final body weights. Apparently, larger and long living fishes allocate more of their energy to somatic growth, whereas the smaller ones to reproductive output. For instance, a 3-fold increase in body size decreases the number of eggs by 20% in apogonids (see Pandian, 2010). From the point of egg and progeny production, aquaculture of smaller fishes as baits for sport fishing and ornamental fishes for commercial purpose may prove several times more profitable.

In the context of maintenance of OSCs, a classical observation by Kaczmarczyk and Koop (2010) in *Drosophila melanogaster* has relevance. In this fruitfly there are early reproducing (S) and late reproducing (L) strains. Reproduction commences in the S strain from the first week of adult life, peaks with the fecundity of 10 eggs/day but completely ceases by the seventh week of adult life. Contrastingly, reproduction in the L strain peaks to produce 10–12 eggs/day during the 7th week of adult life and continues even after the ninth week of adult life. On the whole, the life time fecundity of the early reproducing S strain may not be even a tenth of late reproducing L strain. Incidentally, both the strains commence with 2.5 OSCs/ovariole. However, by the ninth week of adult life, the S strain female has only 0.1 OSC/ovariole against 1.94 OSCs/ovariole in the L strain female. Kaczmarczyk and Koop (2010) have traced the causes for the efficient maintenance of OSCs in late life fertility to a few genes. The signaling pathway of *Dpp*, a member of TGF-β morphogens, plays a central role in maintenance of OSCs. The mutant OSCs, that lack *Dpp* receptors or transcriptional effectors, undergo premature differentiation (Xie and Spralding, 2000). A second TGF-β homologue *Gbb* also contributes to OSCs maintenance. On the opposite side, the cell-autonomous *bag-of-marbles* (*bam*) gene acts on OSCs to promote their differentiation. *Dpp* signaling acts to prevent *bam* expression in the OSCs. The activity of *Dpp* and *Gbb* signals is known to decline with age (Zhao et al., 2008). Apparently, mutation in one or other of these genes may also cause early fertility and reduce the ability to maintain the OSCs, as age advances.

2.14 Stem cells and Germ cells

Embryonic Stem Cells (ESCs) are known to be present in developing embryos and adult tissues of multicellular organisms, except perhaps in pseudocoelomic and haemocoelomic animals (Murugesan et al., 2010). Like the zygotes, the ESCs generate a large number of cells by continuous divisions and ultimately give rise to over 200 different types of cells making up a vertebrate body. The unique features of ESCs namely self-renewal and totipotency provide enormous opportunities to experiment and understand the processes like sex determination and differentiation in fishes, although such experiments are yet to be undertaken. The ES-like cell cultures have been established during recent years, e.g., *O. latipes, D. rerio, Sparus aurata, Lateolabrax japonicus* (see MeiSheng et al., 2010). In India, ES-like stem cell cultures have been successfully developed for the leopard danio *D. frankei* (Routray et al., 2010) and *Lates calcarifer* (Parameswaran et al., 2007). Incidently, Wu et al. (2011) have described a unique epigenetic profile in zebrafish sperm, as has been reported for the mammalian sperm. According to Carrell (2011), the profile in zebrafish sperm includes 'bivalent poising' of key developmental genes of embryogenesis. The bivalency marks a mechanism for establishment of totipotency in the germ cells and maintenance of pluripotency through embryogenesis into the presumptive ES cells. Techniques for generation of fertile diploid nuclear transplants have been successfully developed using embryonic cells as donors (e.g., Wakamatsu et al., 2001).

Generation of ESCs is limited to embryos in mammals and other higher vertebrates but the ESCs are generated in both embryos and adults of fishes or at least until the female reaches reproductive senescence. Hence fishes are unique among vertibrates and provide ample opportunity to identify and isolate the ESCs as well as to manipulate them for a better understanding of the sex differentiation process. Many molecular markers to identify PGCs and trace their migration are known, although only very few molecular markers are now available to identify and techniques to isolate them (Table 22, see also Xu et al., 2010). Germline ovarian stem cells (OSCs) have been identified by Nakamura et al. (2011) in adult ovaries of *O. latipes*. In the 'germinal cradle' located in the stromal compartment of the ovary (Fig. 33), three types of cells buried in the multi-layered epithelium have been detected: 1. Singly isolated germ cells, 2. Cyst forming cells and 3. Large oocytes in the meiotic diplotene stage—all of them surrendered by cells expressing *sox9b*. But the exclusive expression of *nanos* 2 is limited to the singly isolated germ cells, which are the OSCs.

Germ cells: Primordial Germ Cells (PGCs): The PGCs are the progenitors of germ cell lineage and have the ability to differentiate into either oogenia or spermatogonia. As these gonial cells carry heritable information to the next generation, they are considered 'immortal'. Shinomiya et al. (2002) generated chimeras of *O. latipes* by transplanting male (X^1Y^1) and female (X^1X^2) specific PGCs into blastulae committed to differentiate into X^1X^2 and X^1Y^2 genotypes, respectively. From these experiments, they concluded that it is the Germ Cell Supporting Somatic Cells (GCSSCs) that determine the

Table 22. Molecular markers of germ cells of fish

Marker	Activity	Expression and function	Reference
vasa	DEAD box RNA helicase	PGC formation and migration	Yoon et al. (1997), Li et al. (2009)
ziwi	RNA-interacting proteins	Germ cell maintenance	Houwing et al. (2007)
tudors	Tudor domain proteins	Germplasm component, highly conserved across phyla	Strasser et al. (2008)
bucky ball	Novel zebrafish gene	Balbiani's formation	Bontems et al. (2009)
nanos1	RNA binding zinc finger proteins	PGC migration, survival and oocyte maturation	Koprunner et al. (2001), Draper et al. (2007) see also Suzuki and Saga (2008)
nanos2,3	RNA binding zinc finger proteins	Germ cell marker	Aoki et al. (2008)

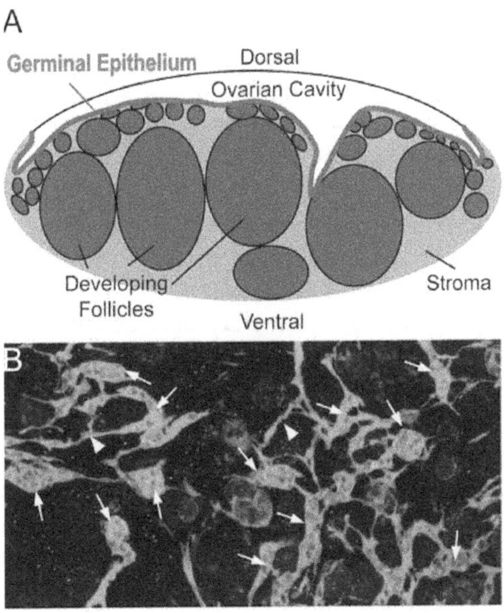

Fig. 33. (A) Transverse section of medaka ovary consisting of (i) ovarian and (ii) stromal compartments. Between these two compartments, note the presence of germinal epithelium lined by the basement membrane. (B) A dorsal view of the ovarian cord in ovary of *sox9b*-EGFP transgenic medaka. 'Germinal cradles' (arrows) composed of germ cells and the *sox9b*-expressing cells are connected to each other by the cellular processes of *sox9b*-expressing cells (arrow heads) to form the network, the 'ovarian cords' (from Nakamura et al., 2011. Ovarian germline stem cells in the teleost fish, medaka (*Oryzias latipes*). Int J Biol Sci, 7: 403–409, with permission of IJBS Publishing Team)

Color image of this figure appears in the color plate section at the end of the book.

sex of medaka progenies. Experimental depletion of PGCs during the early embryonic stage induced the production of male progenies alone in medaka (Slanchev et al., 2005), pearl danio (Saito et al., 2008) and zebrafish (Siegfried and Nusslein-Volhard, 2008). However, the experimental knockdown of *Cxcr4,* the chemo-PGCs attractant receptor gene in medaka by Kurokawa et al. (2007) has shown that the GCSSCs are pre-disposed toward male development and the presence of PGCs is essential for sustenance of sexual dimorphism (Pandian, 2011). During recent years, many publications have demonstrated the critical role played by the maternal genes and their products controlling the early embryonic events including DNA repair and germ cell development (e.g., Abrams and Mullins, 2009; Lindeman and Pelegri, 2010). In teleost fishes, the Yolk Syncytial Layer (YSL) has been considered as extra-embryonic (Sadler, 2004). It is now known that some descendants of each of these tissues are incorporated into the adult zebrafish (Oteiza et al., 2008). Describing a family of 10 genes *zgc,* Hong et al. (2010) have found that about four of them protect the embryo and the others presumably signal the formation of mesoderm and endoderm; indeed Krens et al. (2008) have identified *ERK1* genes as responsible for dorso-ventral patterning and subsequent embryonic cell migration and *ERK2* for cell migration and differentiation, and patterning of mesoderm. In fact maternal gene products continue to play a role in embryonic development even upto organogenesis (Ryu et al., 2005). Many of these recent findings suggest that the germ cell supporting somatic cells in the gonad autonomously activate male genotype but the germ cells activate the differentiation of female phenotype by sending signals to repress male pre-disposition and maintain feminization. Thus it is the 'intimate cross-talk' (Tanaka et al., 2008), and interplay between the PGCs and somatic supporting cells that seem to determine sex and subsequent differentiation (see Pandian, 2011).

It must also be indicated that not all the PGCs in the fish ovary undergo the asymmetric mitosis to produce one of their daughter cells to become the OSC and the other to further undertake cystic division. As the oogenesis advances in embryonic gonad in *O. latipes,* for instance some PGCs, instead of forming the cyst and undergoing meiosis, undergo mitotic division. These PGCs are histologically identifiable by their large size (~ 20 µm in diameter), large nucleus (6–10 µm) and relatively little cytoplasm, i.e., a high nuclear cytoplasmic ratio (see Xu et al., 2010). Germ line specific marker proteins such as nanos, vasa, tudor can also be used as reliable markers (Aoki et al., 2008; see also Table 22). Randomly scattered over the entire embryonic gonad, "these OSCs persist throughout gonadal development, providing an inexhaustible source of cells for oogenesis" (Saito and Tanaka, 2009). Table 23 lists landmark developments in identification, isolation and use of PGCs, SSCs and OSCs to generate allogenics and xenogenics.

Table 23. Land mark achievements in germ cell research on fishes.

Team	Events
Hamaguchi (1982)	PGC migration in medaka
Hong et al. (1996)	ES cell lines from medaka
Olsen et al. (1997) Yoon et al. (1997)	*Vasa* as germ cell marker
Hong et al. (1998)	ES cell-derived chimeras
Yoshizaki et al. (2000)	Transgenic rainbow trout with labelled PGC
Wakamatsu et al. (2001)	ES-derived fertile diploid medaka
Yoshizaki et al. (2002)	Mass isolation of PGCs in rainbow trout
Takeuchi et al. (2003)	Rainbow trout fry by PGC transplantation
Takeuchi et al. (2004)	Xenogenic trout using masu as surrogate
Okutsu et al. (2006)	Allogenic trout by SSCs transplantation
Yoshizaki et al. (2010)	Allogenic trout by OSCs transplantation
Yoshizaki et al. (2010)	Allogenic trout by OSCs transplantation to masu

Incidentally, spermatocysts in the adult testis undergo spermatogenesis in the presence of 11-KT. But, they choose an alternate pathway and form Spermatogonial Stem Cells (SSCs) in the presence of estrogen (see Pandian, 2011). It remains to be seen which steroidogenic hormone induces these PGCs to OSC pathway.

Isolation of germ cells: To isolate the PGCs of the rainbow trout, Yoshizaki et al. (2010) have constructed a transgenic transcript with the *Gfp* gene driven by the *vasa* gene regulatory regions. To reduce the contamination by somatic cells in the partially isolated PGCs, Takeuchi et al. (2003) devised a flow cytometric separation technique to sort out Gfp-positive from Gfp-negative somatic cells. Even with the molecular marker and flow cytometric isolation, the SSCs concentration could be increased to > 56% in the testicular cell suspension used for transplantation. To avoid such contamination, Saito et al. (2008, 2010) and Higaki et al. (2010) have chosen to introduce only one PGC of pearl danio or danio for transplantation.

Two different procedures have been followed to isolate SSCs from adult fishes. Briefly, Lacerda et al. (2006, 2010), and Majhi et al. (2009) lysed the testis of Nile tilalpia and pejerrey, respectively; Lacerda et al. labeled the SSCs by fluorescent cell linker PKH26 and the SSCs were isolated by percol gradient separation. On the other hand, Okutsu et al. (2007) and Yoshizaki et al. (2011) prepared spermatogonial cell suspension of testis from dominant red or white body colored male rainbow trout carrying *pVasa-Gfp* and the isolated labeled SSCs were concentrated flow cytometrically. To obtain pure SSCs, Yoshizaki's team employed two ingenious procedures. They developed (i) molecular markers *rtili*, *rt-scp3* and *rt-shippo* for spermatogonia, spermatocytes and spermatids of rainbow trout, respectively and (ii) the *vase-Gfp* sequence present in the flow cytometrically (FCM) isolated *Gfp-*

positive PGCs expressed abundantly in the gonads but its fluorescence began to diminish with the advancing spermatogonial process. Based on *Gfp* intensities, tubicular cells were divided into a few fractions. Of them *ritili* is used to isolate spermatogonia A type. Lacerda et al. (2006) injected 2 ml of suspension containing $5x10^6$ cells/ml, while Okutsu et al. transplanted only 18,000 cells containing 10,000 SSCs.

OSCs from the dominantly orange body colored transgenic rainbow trout carrying *pVasa-Gfp* were dissociated and concentrated OSCs flow cytometrically were isoloated. From this ovarian cell suspension, 20–30 nl containing 15,000 cells including about 100 OSCs were used for transplantation.

Vasa: The RNA helicase vasa is a germ cell marker in animals and its homologue in vertebrates to date is considered as limited to bisexual reproduction, although it remains unclear with reference to sequential hermaphroditic fishes, which mature first either as females (protogyny) or males (protandry) and subsequently change to the opposite sex. Xu et al. (2005) have cloned *Cag Vasa*, a *Vasa* homologue, from the gibel carp *Carassius auratus gibelio*, a fish known to shift between gono- and gyno modes of reproduction. In bisexually reproducing gibel carp, *vasa* is maternally provided (see Pandian, 2011) and its zygotic expression is restricted to gonads. Following the dynamic sub-cellular distribution, *vasa* RNA is found most significantly in spermatogonia and spermatocytes entering or undergoing meiosis but not in meiotic products namely the spermatids and sperm, as against the continued low-high-low presence throughout oogenesis from oogonia through vitellogenic oocytes to maturing oocytes. The sequence of differential expression of vasa RNA and its protein in oogenesis and spermatogenesis in gibel carp is also common to medaka, tilapia (Kobayashi et al., 2000) and zebrafish (Knaut et al., 2000, 2002). These findings reveal that oogonial cells have retained sexual plasticity throughout oogenesis from oogonia to mature oocytes, whereas it is limited to the primary and secondary spermatocytes alone. Incidentally, a first line normal spermatogonial stems of medaka fish has been established and the line has retained the competence to long-term self-renewal and test-tube sperm production (Hong et al., 2004).

In an attempt to understand the genetic mechanism of differentiation in the gono- and gyno-gibel carp *C. auratus gibelio*, Du et al. (2008) found no expression of *Hira* (histone regulation) transcription in the liver of gyno-carp but a high level of *Hira* RNA in gono-carp. They suggested that the *Hira* may be associated with different modes of reproduction in the carp. Wu et al. (2009) cloned a novel oocyte-specific variant H2A implying a possible association between chromatin structure and reproduction of the gibel carp, which can switch between gynogenic and bisexual modes of reproduction.

The choice between modes of reproduction depends ultimately on the role of germ cells.

Through a series of path-breaking publications, Yoshizaki and his group have demonstrated that PGCs can be transplanted to homologous and heterologous alevins to generate allogenics and xenogenics, respectively. From the point of sex determination and differentiation, PGCs have retained the bisexual potency so long as the recipient of the transplantation is limited to a pre-embryonic (e.g., medaka) and post-hatching stage with undifferentiated gonad (e.g., rainbow trout, Takeuchi et al., 2003) (see Pandian, 2011). For instance, the sex ratio was 0.55 ♀ : 0.45 ♂ for the allogenic rainbow trout *O. mykiss* generated through transplantation of the minimum required numer of (cf Saito et al., 2008) 5–10 PGCs drawn from 30 dpf old dominant orange colored trout into the blastulae of the recessive gray colored trout (Takeuchi et al., 2003). Both male and female xenogenic trout fry were also generated by the recipient *O. masou* blastulae, into which 20-30 PGCs of *O. mykiss* were transplanted (Takeuchi et al., 2004), although 50% xenogenic males and 2.6% xenogenic females alone were fertile (Okutsu et al., 2008).

Avoiding the costs of producing sterilized testis Shimada and Takeda (2008) have used naturally available hybrid steriles as recipients. A cross between the Japanese medaka *O. latipes* ♀ and Chinese Hainan medaka *O.curvinotus* ♂ produce sterile offspring; however, the gonads are sexually differentiated in these hybrids but oogenesis and spermatogenesis are severely impaired (Hamaguchi and Sakaizumi, 1992). Hence the use of sterile hybrids provides an additional advantage by offering sterile ovaries and testes. The main objective of Shimada and Takeda (2008) has been to develop a new method for production of maternal-zygotic medaka but some of their observations are highly relevant to allogenesis.

Orange colored mutant *fgfr1* medaka labeled with rhodamine dextran at one cell stage served as the donor source of the labeled PGCs. Blastulae, generated by the cross between wild gray colored Kaga strain of *O. latipes* females and Hainan *O. curvinotus* males, were the recipients. About a 100 donor cells, drawn from the deeper layer of the blastoderm, where PGCs are known to reside (Kurokawa et al., 2006), were transplanted into the animal pole of the recipient blastula (Fig. 34). Out of 174 transplanted recipients, 52 (i.e., 30%) were found to have the donor-derived PGCs that have successfully migrated to the recipients gonad. Among the orange colored 20 recipient females, fertility was recovered in 12, i.e., 60% (Fig. 34). The allogenic recipient female grew and matured faster than the control and commenced laying eggs at 5–6 weeks of age and produced about 210 eggs each. These females stopped spawning within several weeks and once they stopped, they never produced eggs again. On the other hand, the allogenic males failed to recover fertility, although they matured as

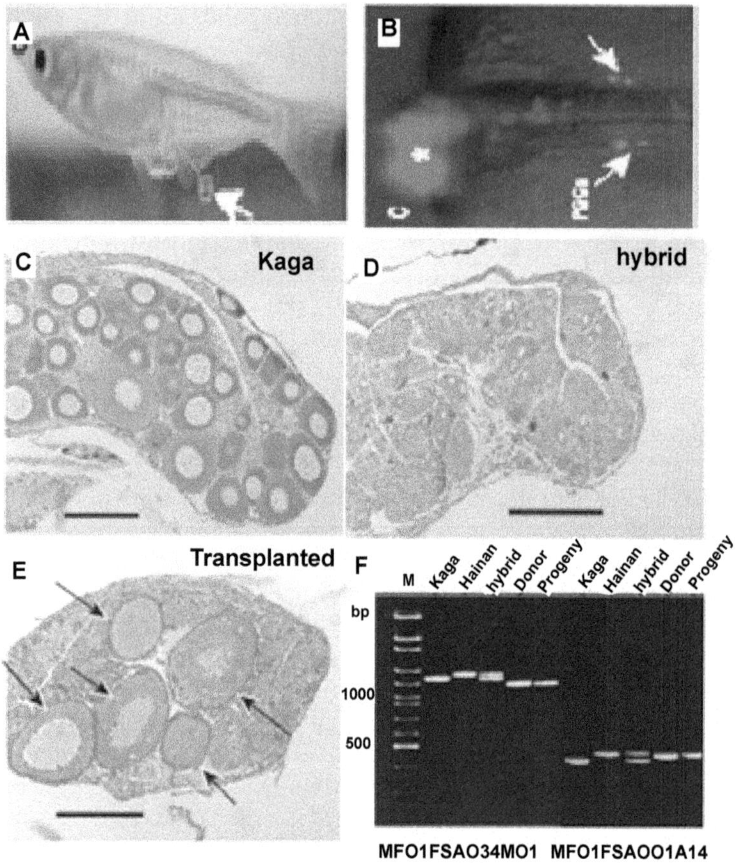

Fig. 34. Recovery of fertility in sterile hybrid medaka by PGCs transplantation from fertile medaka. A = medaka with arrow indicating egg being spawned, B = colonization of 'undifferentiated' sterile gonad by transplanted PGCs, C = Cross sections of ovaries from a 4 week old Kaga strain, D = Non transplanted hybrid, E =Transplanted hybrid ovary contains growing oocytes, as shown by arrows. Scale bar, 200 µm and F = Confirmation of germ-line transfer by detection of strain-specific genetic markers. Note that the patterns of marker amplification for the progeny are identical to those of the donor (from Shimada and Takeda, 2008)

Color image of this figure appears in the color plate section at the end of the book.

males but continued to remain sterile. Notably, the transplantation of PGCs into the recipients blastulae, that have 'undifferentiated' but sterile testes and ovaries, let the PGCs to develop males and females at the ratio of 0.6 : 0.4. An ideal experiment, that is urgently required is the one, in which the PGCs, SSCs or OSCs are transplanted into sterile ovaries of adult hybrid recipient.

OSCs: As indicated by Xu et al. (2005), the PGC-derived OSCs have also retained sexual plasticity. Yoshizaki et al. (2010) intraperitoneally transplanted 20–30 nl of ovarian cell suspension containing ≈ 15,000 cells with 12% *Gfp*-positive germ cells of the trout into 25 dph alevin. At 20 day post-transplantation (dpt), 33 out of 55 recipients showed the presence of donor-derived germ cells (Fig. 35A,A1), and the gonads of each of the *Gfp*-positive recipients were colonized by 3.3 ± 2.6 germ cells. However, on the 150 dpt,

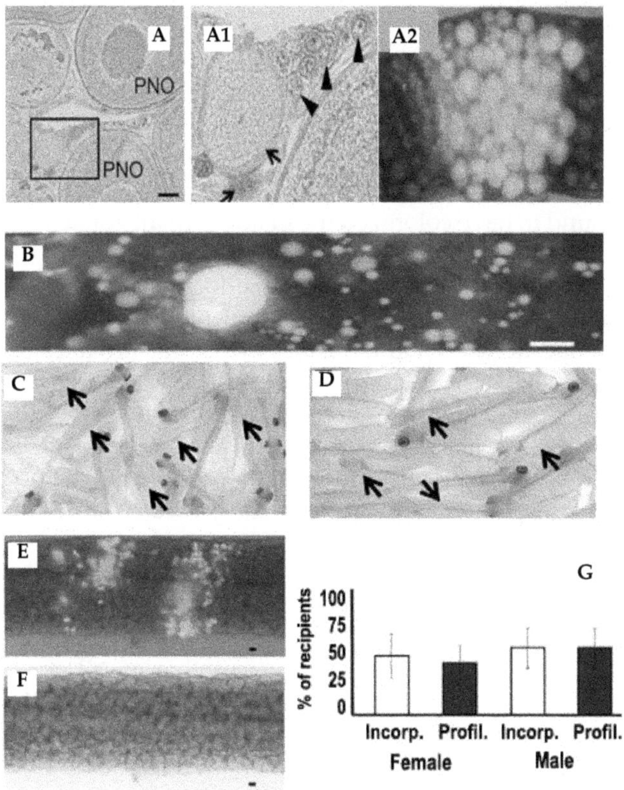

Fig. 35. Colonization of recipient gonads by ovarian germ cells of *Oncorhynchus mykiss*. (A-A_1) perinuclear oocytes (PNO) and oocytes (arrow heads) showing *Gfp*—specific antibody. (A2) *Gfp*-labeled donor germ cells proliferated and differentiated into oocytes in recipient ovary. (B) *Gfp*–labeled donor-derived germ cells remaining in the post-ovulatory follicles of 3n recipient. 50% F_1 progenies were OSCs-derived orange colored females (C) and males (D). Fluorescent (E) and bright-field (F) views of colonization and proliferation of *Gfp*-labeled donor germ cells in the testes of 5 month old recipients. (g) The percentage of progenies carrying the donor-derived germ cells in their gonads (black bars) and those, who did not differ significantly between female recipients and male recipients (from Yoshizaki et al., 2010. Sexual plasticity of ovarian germ cells in rainbow trout. Development, 137 : 1227–1230, doi : 10.1242/dev 044982, reproduced with kind permission by The Company of Biologists Ltd)

Color image of this figure appears in the color plate section at the end of the book.

24 out of 51 showed the presence of 133 ± 110 donor-derived germ cells colonizing each of the undifferentiated gonads of receipients, indicating that competing against endogenous PGCs, the OSCs could colonize only 47% of the recipient's gonads. In a second series, the OSCs, drawn from the dominant orange colored fertile diploid trout, were transplanted into the recessive gray colored sterile triploid masu alevins. After two years of post-transplantation (pt), one mature female out of 20 female recipients, on pairing with normal diploid gray colored trout males, produced all orange colored fry (Fig. 35C,D) and its post-ovulatory ovary contained many oogonia and early oocytes (Fig. 35B), indicating that the donor-derived oogonia drawn from the mitotic ovary can also support oogenesis for multiple spawning during a spawning season in the allogenic triploid recipient ovary.

On being transplanted into presumptive male alevins, the ovarian cells were found to have colonized the male recipient gonads and resumed spermatogenesis (Fig. 35E,F). At the same age 2+, 11 out of 20 triploid male recipients matured. Of them only three produced normal white milt. The partially sterile triploids of many fishes are known to produce aneuploid and euploid sperm (see Pandian, 2011). Hence for progeny testing, only the three completely fertile males were chosen. They sired F_1 progenies, whose sex ratios, namely the orange colored to gray ones, and *pvasa-Gfp*-positive to *pvasa*-Gfp-negative ones, were both nearly 1 : 1. Clearly, OSCs have retained bisexual potency, but are allowed to express only in alevins with an undifferentiated gonad, in which the germ cells supporting somatic cells have also retained bisexual potency.

In the light of the findings by Xu et al. (2005), the successful intraovarian transplantation of stage I follicles in zebrafish becomes interesting. Csenki et al. (2010) separated stage I follicles from the ovary of YFP strain and injected them into the ovary of the foster mother of AB strain (Fig. 36). In the foster mother, stage I follicles developed into stage IV within 14 days, although the number of surviving transplanted follicles decreased from 47.5% at 1 dpt (day post-transplantation) to 2.6% at 14 dpt, perhaps owing to immunological incompatabilities. Still, the development of four YFP embryos from two foster mothers was followed. Microsatellite analysis of eight independent genomic loci clearly demonstrated that these transplanted oocytes successfully developed into fertilizable eggs and on fertilization, produced viable offspring (Fig. 36). From the point of sex differentiation, the unisexual potency of the germ cells supporting somatic cells and hormonal micro-environment in the adult ovary of the foster mother could only facilitate the transplanted alien ovarian follicles to develop into eggs alone.

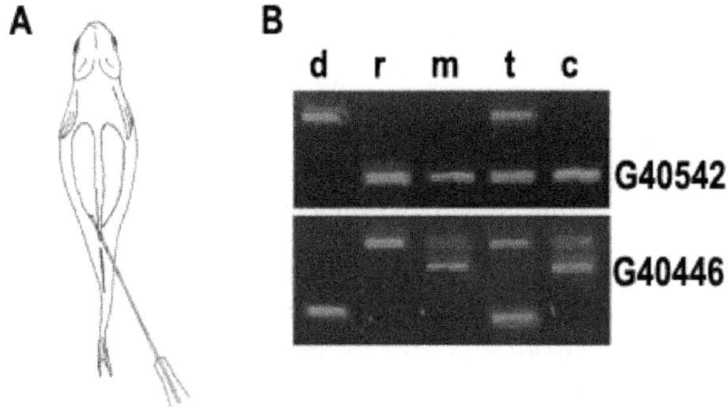

Fig. 36. Transplanted zebrafish follicles invade the recipient ovary and successfully continue oocyte development. (A): Schematic picture of the process of transplantation. (B): Microsatellite analysis of genetic composition of (d) donor YFP female, (r) recipient AB female, (m) male used for crossing with recipient female, (t) offspring developing from transplanted follicle and (c) sibling from recipient female (from Csenki et al. 2010, reproduced with permission by The International Journal of Developmental Biology, 54: 585–589)

Following Shimada and Takeda (2008), Wong et al. (2010) have also chosen to use the hybrid danios to produce zebrafish germline chimeras using hybrid sterile fry as recipients but of germ cells that have been isolated from the adult ovary for transplantation. Like the hybrid medaka fish, the hybrids between pearl danio *D. albolineatus* ♂ and zebrafish *D. rerio* ♀ are sterile. The male hybrid has no spermatid and spermatozoa but everything of a testis; similarly, the female hybrid has no oocytes but everything of an ovary. The donor OSCs were obtained from the ovary of 3-month old adult Tg (*vasa: DsRed 2-vasa*): Tg (*bactin: EGFP*) double transgenic zebrafish that express *EGFP* throughout the body under the control of *bactin* promoter and *DsRed* specifically in the germ cells driven by the *vasa* promoter. Using this strategy, it was possible to visualize the true allogenics as well as the transplanted OSCs. To obtain the OSCs, the ovary was disassociated and the cells fractionated on a discontinuous Percoll gradient, which increased the ratio of *DsRed*-positive and *EGFP*-positive cells from 8.9 in the initial cell suspension to 193 per 10,000 cells after Percoll fractionation. About 755 *DsRed*-positive OSCs were directly transplanted into the abdominal cavity of a two week-old hybrid fry.

Five days following the OSCs transplantation, 2.8 *DsRed*-positive cells were located in each of the 104 recipients. Eight weeks after transplantation, the transplanted OSCs colonized two out of 10 recipients gonad. Of 67 adult recipients, there were nine females and 12 males and in the others fertility was not recovered. All the females were also sterile, as they did not breed with male zebrafish or pearl danio. Of 12 males, six were fertile and fertilized wild female of zebrafish and produced F_1 progenies at the sex ratio of 0.43

♂ : 0.57 ♀. The results of Wong et al. (2010) confirm that the OSCs drawn from the ovary of 3-month old zebrafish still retain bisexual potency and the germ cells supporting somatic cells of undifferentiated gonad in the two week-old hybrid fry have also retained bisexual potency. For reasons not known, the fertility is not recovered, when hybrid sterile fry are used as recipients in male medaka progenies (Shimada and Takeda, 2008) but in female danio progenies.

SSCs: The PGC-derived SSCs have also retained sexual plasticity. Okutsu et al. (2007) intraperitoneally microinjected testicular cell suspension containing 18,000 cells with 1,000, i.e., 56% labeled SSCs (as against 12% labeled OSCs in Yoshizaki et al., 2010) drawn from the dominant orange colored diploid rainbow trout into the alevins of triploid sterile masu salmon. The resultant xenogenic trouts were both males and females, clearly indicating that bisexual potency has been retained not only by the SSCs but also the germ cells supporting somatic cells in the undifferentiated gonads of the alevins. On maturation, 84% of the recipient males produced milt and 50% of the sired F_1 progenies, which developed the finger print pattern like that of their fathers. Similarly, 10% of the recipient females matured and 90% of their eggs hatched.

In another series of experiments, Yoshizaki et al. (2011) transplanted testicular cell suspension of the dominant orange colored diploid rainbow trout into the alevins of triploid masu salmon. Two years after the transplantation, the recipients showed nuptial body coloration and males with extended jaws, both of which are typical secondary sexual characteristics of salmonids. In yet another experimental series, Yoshizaki et al. transplanted 500,000 cryopreserved and thawed testicular cells into the alevins of 3n masu. At the second spawning season, 10 mature males and five females of masu salmon produced the donor-derived diploid and aneuploid sperm. On the other hand, F_1 female progenies produced functional eggs. Hence it is possible to generate functional eggs and sperm using the cryopreserved SSCs of the trout, so long as the recipients are alevins that have undifferentiated gonad with germ cells supporting somatic cells still retaining bisexual potency.

In the early 90s, another Japanese group led by CA Strussmann introduced the sperm cells into sterile testis of the pejerrey *Odontesthes bonariensis* and found the sperm to have been alive until 10 week pt (see also Pandian, 2003). Recently his group surgically transplanted the donor spermatogonial cells, isolated from juvenile pejerrey into the sterile adult Patagonian pejerrey *O. hatcheri*, which have been previously partially depleted of endogenous gonial cells by a combination of treatment with busulfan (40 mg/kg) at high water temperature (25°C). The donor-derived SSCs have colonized the recipient's gonad and resumed spermatogenesis within 6 months (Fig. 37). The presence of donor-derived gametes has been confirmed by PCR in 20% recipient males of Patagonian pejerrey. On crossing these xenogenic males with pejerrey females, the progenies are either hybrids or pure pejerreys with donor-derived germline at the frequencies of 1.2 to 13.3%. Hence the transplantations of SSCs into the sterile adult male allogenic *O. niloticus* and that of *O. bonariensis* into sterile adult male xenogenic *O. hatcheri* have resulted in the production of males only, i.e., apparently, the germ cells supporting somatic cells have lost bisexual potency and the hormonal micro-climate prevalent in the adult sterile testis allow the SSCs with bisexual potency to express unisexual potency, i.e., develop testis alone.

In this context, a publication by Tagami et al. (2007) becomes relevant. They have transplanted 'W' bearing PGCs into male recipient embryos, and the female PGCs have colonized the recipient's gonad. In the 'testis' of allogenic the female PGCs have been differentiated into spermatogonia (30.8%), spermatocytes (32.7%), spermatids (28.4%) and spermatozoa (0.2%). Hence the female PGCs in the testis of the allogenic cock successfully pass through the first and second meiotic divisions but are hardly capable of spermeiogenesis.

While the technical feasibility of allogenesis and xenogenesis and their potentials are of great significance, research in this area demands very sophisticated techniques and high skills for cell transplantation. Further, the xenogenics may take a considerably long time to mature and undertaking progeny testis. As an alternative, the Brazilian group led by C Franca used a sterile male as a recipient. Lacerda et al. (2006) have claimed that their allogenics of *O. niloticus* have developed testes from the donor-derived SSCs, *albeit* without having undertaken progeny testing. In a subsequent experiment, the SSCs, isolated from red Nile tilapia were transplanted into the busulfan-treated adult testis of gray colored Chitralada tilapia (Lacerda et al., 2010). Fluoresence microscopic evidence showed that the transferred donor germ cells efficiently colonized the testis and generated PKH26-labeled spermatogenic cysts and ultimately formed mature spermatids and spermatozoa. DNA microsatellite analysis revealed the presence of donor-derived alleles in 6.3% of the F_1 progeny (Fig. 37).

The results obtained in all these experiments seem to indicate the following: 1. Competing against the respective endogenous PGCs, the colonizing efficiency of the donor-derived OSCs in allogenic trout is about 41–47%. Among the mature allogenics, 5% females were fertile, as against 15% fertile males. But the fertility of the xenogenic trout was as high as 84% for males and 10% for females. It is not known whether the competence to differentiate into fertile ovary by both OSCs and SSCs is significantly lower than that to differentiate fertile testis in allogenic as well as xenogenics, or

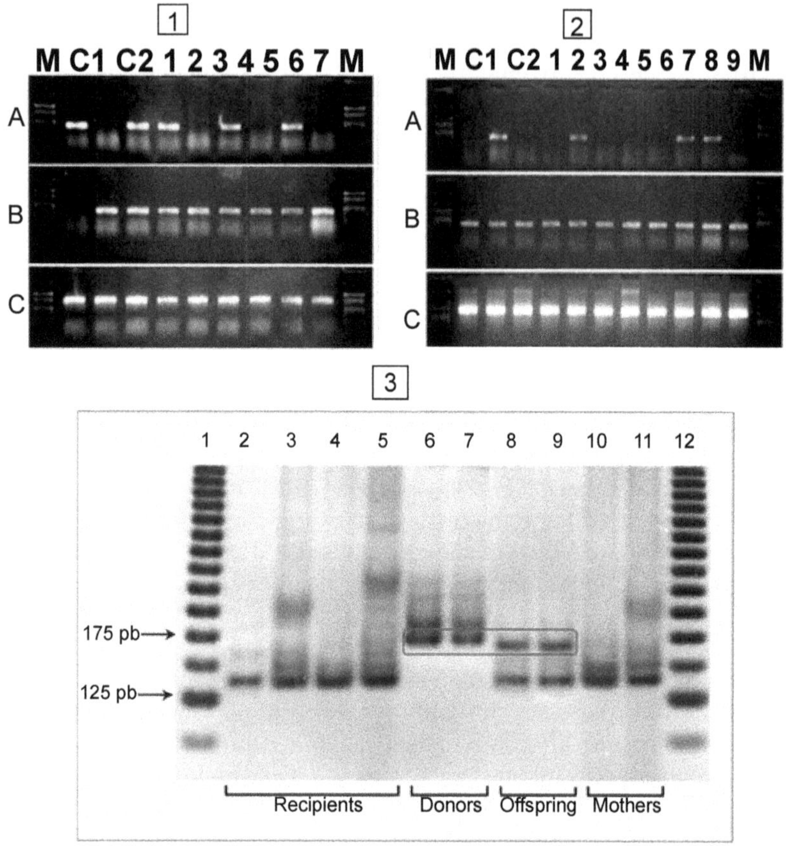

Fig. 37. (1): PCR analysis of sperm of 8-month old recipient *Odontesthes hatcheri*. (A) The donor *O. bonariensis* specific sequence, (B) *O. hatcheri* specific sequence, (C) β-actin control. Note the donor-derived *O. bonariensis* spermatozoa in four recipients, as shown in lanes: 1, 2, 4 and 6. (2): PCR analysis of progenies sired by allogenic *O. bonariensis*/surrogate crossed with normal *O. bonariensis* female (Note the donor-derived *O. bonariensis* spermatozoa in four recipients, as shown in lanes: C_2, 3, 7 and 8) (from Majhi et al., 2009). (3): Genomic DNA analysis of *Oreochromis niloticus*. Alleles from donors were detected in two progenies (rectangle), indicating that these fish were derived from donor cells and were not related to the surrogate recipients (from Lacerda et al. 2010)

these allogenic/xenogenic females suffered greater mortality (Table 24). Incidentally, a reason for the wide differences between the allogenics and xenogenics is that the cell suspension used for transplantation contained 56% gonial cells in the xenogenics, against 12% only in allogenics, and 2. Both OSCs and SSCs have retained bisexual potency even after maturation and milting/spawning by the respective donors but are allowed to express their bisexual potency, so long as the recipients are alevins with sexually undifferentiated gonad, i.e., gonadal differentiation leads to the reduction of bisexual potency of germ cells supporting somatic cells into unisexual

Table 24. Sex ratio of allogenic and xenognic fishes generated by transplantation of PGCs, SSCs or OSCs into the recipients blastulae/alevins or sterilized adult males

Donor (D)/Rccipient (R)	Graft	Sex ratio ♀ : ♂	Reference
	PGCs to early embryos		
Oncorhynchus mykiss (D,R)	15–30 PGCs	0.15:0.85	Takeuchi et al. (2001)
O. mykiss (D,R)		0.55:0.45	Takeuchi et al. (2002)
Carassius auratus (D) *Cyprinus carpio* (R)	'Sandwich'	0.40:0.60	Yamaha et al. (2003)
Danio albolineatus (D) *D. rerio* (R)	One PGC	0.00:1.00	Saito et al. (2008)
C. auratus (D) *D. rerio* (R)	One PGC	0.00:1.00	Saito et al. (2008)
Misgurnus angiullicaudatus (D) *D. rerio* (R)	One PGC	0.00:1.00	Saito et al. (2008)
D. rerio (D) *D. rerio* (R)	One PGC preserved	0.17:0.83	Higaki et al. (2010)
Oryzias latipes (D) Hybrid *O. latipes* ♀ X *O. curvinotus* ♂(R)	100 PGCs	0.4:0.6	Shimada and Takeda (2008)
	SSCs to alevins		
2n *O. mykiss* (D) 3n *O. masou* (R)	10,000 SSC	0.43:0.54	Okutsu et al. (2006)
	OSCs to alevins/blastulae*		
2n *O. mykiss* (D) 3n *O. masou* (R)	100 OSC	0.33:0.67	Yoshzaki et al. (2010)
Danio rerio (D)* Hybrid *D. rerio* ♀ X *D. albolineatus* ♂ (R)*	755 OSCs	0.57:0.43	Wong et al. (2010)
	SSCs to sterilized males		
Oreochromis niloticus (D) *O. niloticus* (R)	SSCs	0.00:1.00	Lacerda et al. (2006)
O. niloticus red (D) *O. niloticus* chitralada (R)	SSCs	0.00:1.00	Lacerda et al. (2010)
Odontesthes bonariensis (D) *O. hatcheri* (R)	SSCs	0.00:1.00	Majhi et al. (2009)
2n *O. mykiss* (D) 3n *O. masou* (R)	500,000 preserved SSCs		Yoshizaki et al. (2000)

potency. Incidentally, it implies that gonadal differentiation means reduction in bisexual potency into unisexual potency in germ cells supporting somatic cells.

Incidentally, dietary administration of androgen to masculinize gravid female with differentiated ovary and gestating embryos with undifferentiated gonad provides unique opportunity to test whether or not the gonadal differentiation induces the loss of bisexual potency of germ cells supporting somatic cells. Kavumpurath and Pandian (1992, 1993) administered selected doses of natural or synthetic androgens to gravid female *Poecilia reticulata* during the labile period (of the gestating embryos) of 5–10 days prior to parturition. They have reported the masculinization of the progenies but not the females. From another series involving the dietary administration of a series of androgens to gravid *P. sphenops*, Geroge and Pandian (1995, 1998) confirmed that none of the treated adult females that received a relatively high dose of androgen, ever reversed to male. These observations clearly show that the process of differentiation induced the loss of bisexual potency of the germ cells supporting somatic cells.

Nevertheless, these findings do not provide a solution to the classical problem of the regeneration of testes in ovariectomized *Betta splendens*. Hence research on transplantations of SSCs into sterile adult female and OSCs into sterile adult male may provide greater insights into this problem of sexual differentiation induced by OSCs/SSCs. The cyprinodontid tooth carps, known for semelparity and the shortest generation time (matures in 6 weeks, see Pandian, 2010) and life span, *Nothobranchius rachovi* and *N. furzeri* (Valdesalici and Cellarino, 2003) may serve as ideal experimental fish. Secondly, reward winning experiments may also be undertaken in the air-breathing anabantid *Macropoclus opercularis*, an iteroparous, frequent spawner with a short life span and a gonochore, known for its ability to change sex in either direction (see Pandian, 2011) shall prove to be an excellent choice. In fact a host of anabantids *B. splendens* and *Trichogaster* are also suggested to have the ability to change sex in either direction (Svardson and Wickbom, 1942). The findings of Tagami et al. recall partially the results reported by Shinomiya et al. (2002). Incidentally, the PGCs are of maternal origin and may have the print of XX genotype. In fish, hitherto generated allogenics and xenogenics involve *O. mykiss*, *O. masou*, *D. albolineatus*, *D. rerio*, *C. auratus*, *Misgurnus anguillicaudatus* (Yasui et al., 2011), *O. niloticus*, *O. bonariensis* and *O. hatcheri*—all of them are male heterogametics. As Farlora et al. (2009) have established a model system for germ cell transplantation in cichlids, it will be interesting to transplant the PGCs from the female heterogametic *O. aureus* into male heterogametic *O. niloticus* and vice versa. The results of these experiments shall provide a key to understand the genetic mechanism of sex differentiation in primary gonochores as well as secondary gonochores and sequential/serial hermaphrodites.

3

Differentiation in Hermaphrodites

Hermaphroditism is defined as the expression of both male and female reproductive function in a single individual either simultaneously or sequentially (Sadovy de Mitcheson and Liu, 2008). In fishes three patterns of functional hermaphroditism have been recognized (Pandian, 2010): simultaneous, sequential and serial (Fig. 2). Simultaneous hermaphrodites function as male and female at the same time or within a short span of time. They do not undergo natural sex change. However, the sequential and serial hermaphrodites undergo natural sex change; the sequentials change sex only once in a single direction (protogyny: female to male; protandry: male to female) during their life time but the serials do it more than once in either direction. The ontogenetic pathways of sexual differentiation in these sex changing sequentials and serials are depicted in Fig. 38. Incidentally, it is known that a fraction of individuals in sex changing sequential populations does not undergo sex change: e.g., protogynics *Pagrus pagrus*, protandric, *Diplodus vulgaris*. Social pressure induced sequential (e.g., Cichlidae, *Crenicara punctulata*, Zupanc, 1985) and serial (e.g., *Macropodus opercularis*, see Pandian, 2011) sex changes do occur in a few gonochores. They are discussed elsewhere (Pandian et al., 2012).

3.1 Patterns of hermaphroditism

Due to plasticity in sexuality and flexibility in mating system, the detection of truly functional hermaphoroditism in some 350–400 species belonging to 34 or so families (Nakazona and Kuwamura, 1987) has been indeed a task (see Pandian, 2010). By applying a rigorous set of criteria to

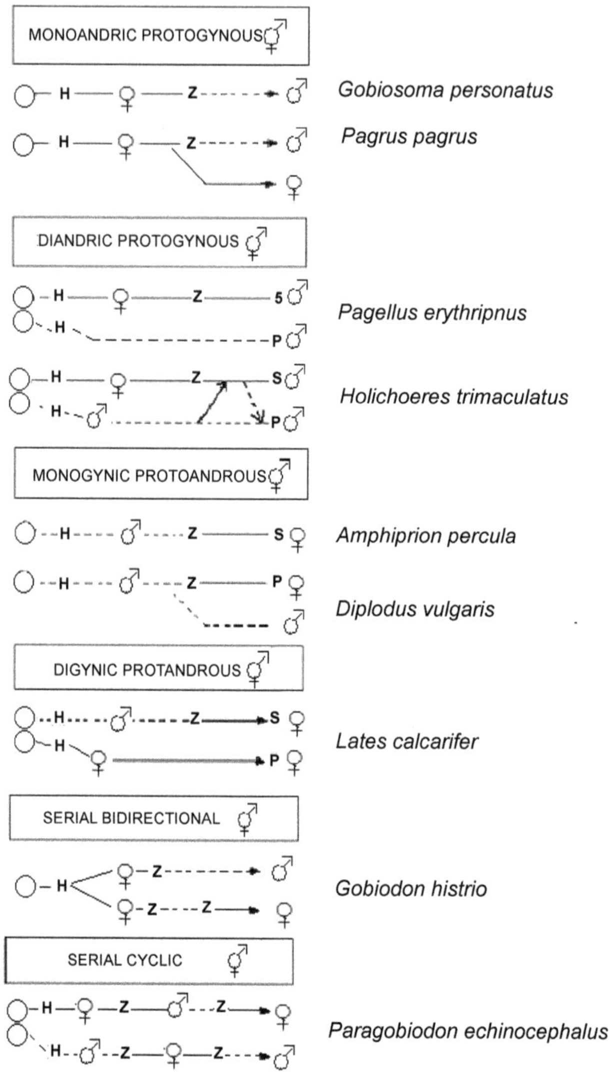

Fig. 38. Ontogenetic pathways of sex differentiation in sex changing hermaphroditic fishes

data gathered from over 400 primary studies, Sadovy de Mitcheson and Liu (2008) have confirmed the existence of functional hermaphroditism only in 137 species belonging to 27 families in 94 genera; for an additional 21 families in 31 genera, confirmation for the existence of functional hermaphroditism is awaited for want of adequate evidence. Since 2008, functional hermaphroditism in more and more fish species is continuously being confirmed, for example diandric protogyny in *Epinephelus coioides*

(Liu and Sadovy de Mitcheson, 2009). Within the 27 families, in which functional hermaphroditism is confirmed, it occurs in at least 6% of all these families and more than 10% of families among the Perciformes. Nevertheless, there are only 123 fish species in which natural sex change occurs once or more than once during their life time; yet they constitute a biologically fascinating group (Pandian, 2010) and commercially important resource (Chopelet et al., 2009). Among these 123 species (Table 25), 74% of them are protogynous hermaphrodites, clearly indicating that this pattern

Table 25. Taxonomic distribution of different patterns of functional hermaphroditism in fishes, in which the indicated pattern of functional hermaphroditism is confirmed or suspected*/almost confirmed (compiled from Sadovy de Mitcheson and Liu 2008)

Order	Simultaneous (no)	Protogyny (no)	Protoandry (no)	Serial (no)
Anguilliformes Muraenidae	1^+		3^*	-
Stomiformes	-	-	$1^+, 2^*$	-
Aulopiformes	$7^+, 2^*$	-	-	-
Cyprinodontiformes	$1^+, 1^*$	-	-	-
Synbranchiformes	-	3^+	-	-
Scorpaeniformes	-	3^+	5^+	1^*
Perciformes	$5^+, 5^*$	$71^+, 17^*$	$19^+, 2^*$	$7^+, 4^*$
Total	$14^+, 8^*$	$77^+, 17^*$	$25^+, 7^*$	$7^+, 5^*$

of sequential hermaphroditism has been the choice of natural selection; the choice of natural selection decreases in the order of : < protandry (20%) < serial (6%). The geographical distribution of these hermaphrodites is limited mostly to the tropical coral reef ecosystem. Except for a few species of *Amphiprion* that have been successfully cultured on a commercial scale in a few countries (e.g., Ajith Kumar and Balasubramanian, 2009), the sequentials and serials are not easily amenable to culture and/or maintained in the laboratory. Consequently, literature available on their sex determination and differentiation is far less, in comparison to that for gonochores. Hence the description on the amazing diversity of these hermaphrodites is likely to be more biological, and shall serve to introduce them to molecular biologists in search of sex determining and differentiation genes.

3.2 Simultaneous hermaphrodites

By number of species, the simultaneous hermaphrodites are a few but in terms of diversity, they fall under three patterns: 1. Self-fertilizing hermaphrodites, 2. Gamete exchange hermaphrodites, 3a. Marian hermaphrodites and 3b. Okinawan hermaphrodites (Fig. 39). Self-fertilizing hermaphroditism, a rarity among vertebrates, occurs in rivulus *Kryptolabias marmoratus* and possibly in *K. ocellatus* (see Pandian, 2011); in them 98 and 97% self fertilization frequencies have been recorded (Tatarenkov et al., 2009), leaving the 2–3% fertilization frequency to secondary males. Harrington (1967) has reported the appearance of < 5% primary males, when the eggs are incubated at < 19°C (Fig. 48). However, exposure of post-hatchlings to 28°C generates secondary males. Relatively higher water temperatures prevailing in tropical Belize produce nearly 25% secondary males (Tatarenkov et al., 2009). Understandably, sex change can be induced in this self fertilizing hermaphrodite by exposing to different thermal regimes during the early

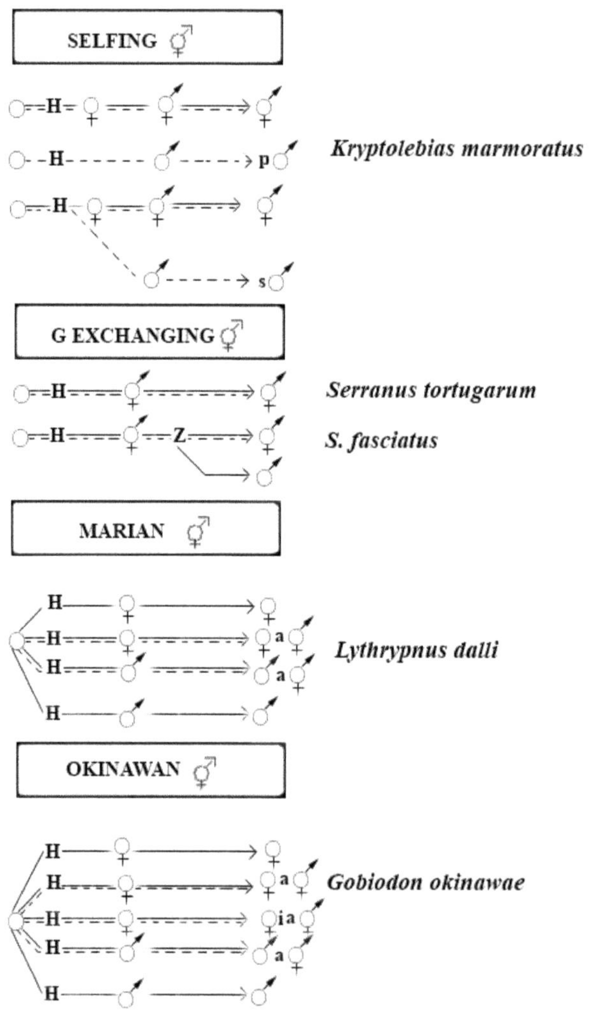

Fig. 39. Ontogenetic pathways of sex differentiation in simultaneous hermaphroditic fishes. H = hatching, G = Gamete exchanging, Z = sex changing, p♂ = primary male, s♂ = secondary male, ♀a⚥= female active hermaphrodite, ♂a⚥ = male active hermaphrodite

labile embryonic and post-hatchling stages only. The juvenile rivulus are also amenable to 100% feminization (Orlando et al., 2006). These field and experimental observations clearly indicate that the hermaphrodite has retained its ability to function as male or female for the last 200,000 years (Tatarentov et al., 2009), and has successfully escaped the Muller's ratchet. Their escape from the Muller's ratchet may be traced to the unprecendented genetic diversity of 1.4 individuals/clones observed among the rivulus population (Turner et al., 1992). To trace the source of such genetic diversity, Nakamura Y et al. (2008) have induced 31 artificial inseminations between two clonal strains characterzed by homozygosity and hererozygosity and found one heterozygous rivulus out of every 13 individuals.

In this context, a publication by Zarkower (2006) becomes relevant. In the nematode *Caenorhabditis elegans*, there are two natural sexes namely XX hermaphrodites and XO males. Hermaphrodites are somatic females but can reproduce either by self-fertilization or by mating with males. Matings generate broods of 50% XX and 50% XO progenies. Selfing progenies are mostly XX but one out of 500 is a XO male, which is generated due to non-disjunction of X chromosomes. Whether a similar non-disjunction of X chromosomes is induced by temperatures below 19°C resulting in the production of 5% (i.e., one out of 20 embryos) primary males in *K. marmoratus* remains to be tested.

Gamete exchanging hermaphrodites: The existence of gamete exchanging hermaphrodites has been considered to have been limited to all species belonging to the two genera *Serranus* and *Hypoplectrus* of Serranidae (Pandian, 2010). Sadovy de Mitcheson and Liu (2008) have recently confirmed its existence in another dozen species belonging to seven families (including Serranidae) in 12 genera. In them, the ovarian and testicular tissues are separated (by a thin basement membrane e.g., *Diplectrum formosum*, Bubley and Pashuk, 2010) and with their respective ducts. Based on the mode of gamete-exchange, there are three groups within this hermaphroditism. The territorial permanent monogamous (e.g., *Serranus tortugarum*) and non-territorial serial monogamous (e.g., *S. tigrinus*) hermaphrodites mate only as male or female during any specific spawning but they have the ability to switch sex roles between successive spawnings. In the third group, an accessory structure composed of epithelial cells that form highly convoluted villi (Fig. 40) is located posterior to the ovotestis; hydrated oocytes, ovulated once every two days during the spawning period of 337 days from March to January, are stored to be spawned when an opposite partner becomes available. The survey of Bubley and Pashuk (2010) has found 22% individuals of *Diplectrum formosum* with stored oocytes in the accessory structure in coastal waters from North Carolina to Florida, USA.

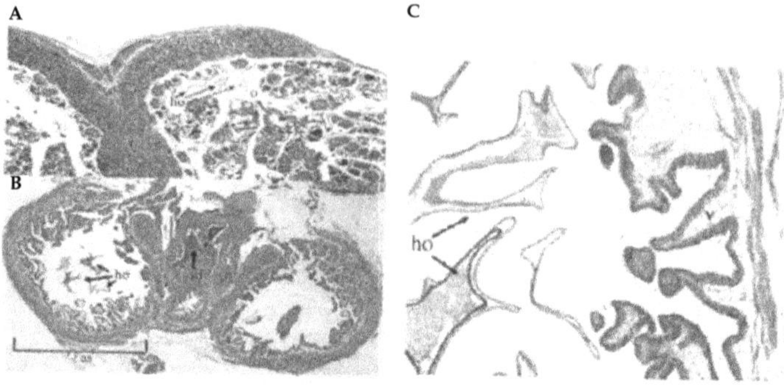

Fig. 40. Upper panel: Transverse cross-sections of mature ovotesis of *Diplectrum formosum* showing compartmentalization (A) t = ripe testicular and o = ovarian tissues and (C) between sd = sperm duct with spermatozoa and as = accessory structure. Ho = Hydrated oocytes are present within the ovarian portion (x20). Lower panel: Villi-like projections (v) from the wall of the accessory structure in the ovotestis with the presence of hydrated oocytes (ho) (x400) (from Bubley and Pashuk, 2010). Life history of a simultaneous hermaphroditic fish *Diplectrum formosum*. J Fish Biol, 77: 676–691 (reproduced by permission of The Fisheries Society of the British Isles/John Wiley & Sons Ltd)

A combination of low density, limitation of space, reduction in allocation for female function to 56% from 77% and shortened spawning period has driven a few gamete exchanging simultaneous hermaphrodites to switch over to pure male (Fig. 39). Thus, all individuals of *S. baldwini* and *S. fasciatus* commence life as simultaneous hermaphrodites but the largest individuals in any location lose the female function and become functional males; these males hold territories and harems consisting of one to seven females.

Marian hermaphrodites: Thanks to the series of publications by St. Mary, the existence of Marian hermaphrotism in about five goby species belonging to genus *Lythrypnus* has been established (Fig. 2). Marian hermaphrodites have simultaneously different fractions of both ovarian and testicular tissues in their gonad, as in simultaneous hermaphrodites but nonetheless function either as male or female at any point of time. Unlike the Okinawan hermaphrodites, the Marians are solitary. Depending upon the quantum of tissues present in the gonad, there can be pure males, male-biased hermaphrodites, female-biased hermaphrodites and pure females. In *L. zebra, L. spilus* and *L. phorellus*, pure males are absent but their function is carried by male biased hermaphrodites. According to St. Mary, they may be capable of bidirectional sex change but successful sex change is described in *L. zebra* only (see Pandian, 2010).

Okinawan hermaphroditism has been described in two species *Gobiodon okinawae* from the Lizard Island (Cole and Hoese, 2001) and *Dascyllus aruanus* from Papua New Guinea (Cole, 2002). They more or less resemble Marian hermaphrodites but are colonials. In some of their colonies, a male is absent and a male-active hermaphrodite takes over the function of the male. In both Marian and Okinawan hermaphroditism, the male ratio is progressively reduced to almost zero.

3.3 Sex changing hermaphrodites

In fish sex change is a unique but an intriguing phenomenon, especially from the point of genetic basis of sex determination. The presence of PGCs has been reported from differenitiated ovotestes of *S. hepatus*, *Coris julis* (Brusle, 1988), *Epinephelus microdon* and *A. frenatus* (Brusle-Sigard, et al., 1992, 1994; Reinboth and Brusle-Sigard, 1997), *A. polymnus* (Rattaunayuvakorn et al., 2006), *E. coioides* (Yao et al., 2007) and *H. trimaculatus* (Kojima et al., 2008). Similar reports on the presence of PGCs in ovotestes/testis of other sex changing fish are awaited. The simultaneous presence of PGCs, SSCs and OSCs has been detected in the ovotestis of *A. frenatus* five weeks following the commencement of sex change from male to female. However, the sex-change is attributed to the PGC-derived spermatogonial and oogonial germ cells rather than the PGCs themselves (Brusle-Sigard et al., 1994). By the division and differentiation into spermatogonia in testis of *C. julis*, the intragonadal PGCs do play an active role (Reinboth and Brusle-Sigard, 1997). Transplantation of PGCs or their derivatives OSCs or SSCs to homologous/heterologous recipient embryos with undifferentiated gonad has been shown to differentate into males and females . When the recipient is an adult with already differentiated but sterilized testes containing germ cell supporting somatic cells alone (see Lacerda et al., 2006), the transplanted SSCs result in the differentiation of testes only (Lacerda et al, 2006, 2010; Majhi et al., 2009). It is not yet known whether the PGCs and/or the germ cell supporting somatic cells play a major role in induction and realization of sex change.

In long-living seasonal spawnering primary gonochores, a major fraction of germ cells component is degenerated following the completion of single or multiple spawnings. Gonadal recrudescence is synchronized with the onset of the ensuing pre-spawning season. However, the germ cell component in these adults differentiates into only spermatocytes in the testis or oocytes in the ovary. Hence the germ cell supporting somatic cell component, which remained intact (Fig. 41), seems to direct the gonadal differentiation in the direction of the original genetic sex.

In secondary gonochores such as *D. rerio* and *A. anguilla*, the majority of PGCs initially differentiate into oocytes. However, on apoptosis of the oocytes and ovarian cavity in a short period of time, 10 days in zebrafish, in the presence of the same germ cells supporting somatic cells about 50% of the presumptive males are transformed into functional males. Thus both germ cells and germ cells supporting somatic cells in the differentiated but non-functional ovaries have still retained bisexual potency (see also Nakumara et al.,

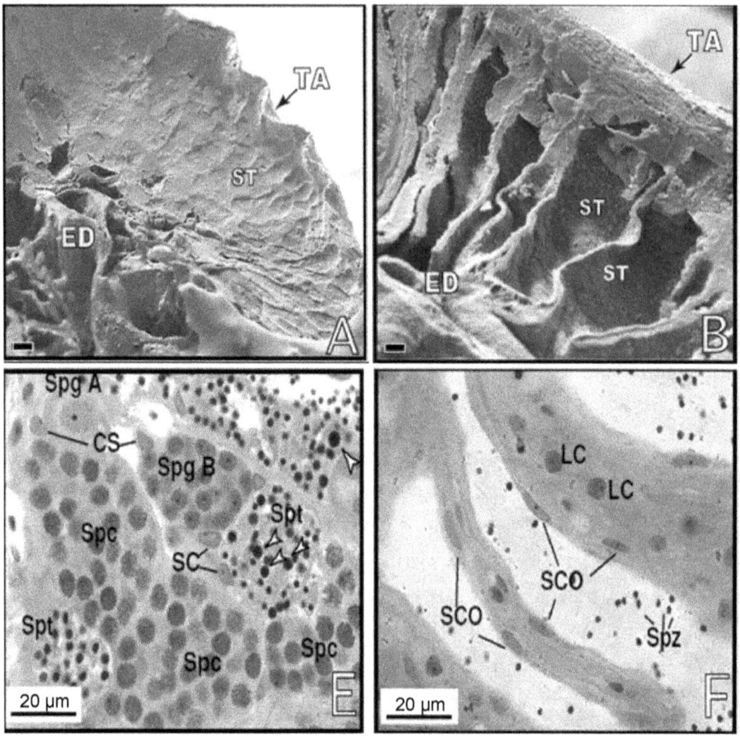

Fig. 41. Transverse sections of the matured testis of *Oreochromis niloticus*. A and B are scanning electron microscopic views of (A) fertile and (B) sterile testes. Light microscopic views showing the spermatogonial cysts in different phases of development in (E) fertile and (F) sterile testes. TA = tunica albuginea, ED = efferent tuctules, ST = seminiferous tubules, spg A and B = types of spermatogonia, spc = spermatocytes, spt = spermatids, SC = Sertoli cells, LC = Leydic cells. SCO = Sertoli cells only, Spz = sperm (from Lacerda et al., 2006)

2003). Further, there are also other similarities; for instance, *Dmrt1* expression is limited only to spermatogonia and spermatocytes in both the secondary gonochoric zebrafish (Guo et al., 2005) and protogynous sequential *E. akaara* (Xia et al., 2007). It is likely then that in the protogynous hermaphrodites too, the transformation of 'adult sex change' is facilitated by the bipotential germ cell supporting somatic cells. It remains to be seen whether this pattern of 'juvenile sex change' observed among the secondary gonochores is also in operation in 'adult sex change' of protogynous hermaphrodites.

The damselfish genus *Dascyllus* (Pomacentridae) comprises of both gonochoric and protogynous species and provides an excellent system for a study of evolution of protogyny (McCafferty et al., 2002). Not surprisingly, a series of studies have shown the existence of gonochorism in *D. trimaculatus* (Asoh and Kasuya, 2002) and *D. albisella* (Asoh et al., 2001) but protogyny in *D. aruanus* (Cole, 2002), *D. reticulatus* (Schwarz and

Smith, 1990), *D. marginatus* (Shpigel and Fishelson, 1986) and *D. carneus* (Asoh and Yoshikawa, 2003). The sequence of gonadal differentiation in *D. flavicaudus* proceeds from ovarian stage I to stage II and to the formation of ovarian lumen, a stage corresponding to non-vitellogenic peri-nuclear stage in zebrafish (Wang et al., 2007). More or less the same sequence of gonadal differentiation has also been reported for *D. albisella, D. trimaculatus, D. aruanus* and *D. carneus*. From a detailed study on the distribution of mixed stage (consisting of both ovary and testis at different stages) and other gonadal categories with increasing size of *A. flavicaudus* from the northern and southern Vaipahu outer-reef populations in Moorea of French Polynesia, Asoh (2004) has come to the following conclusions: 1. Beyond the ovarian stage II, *D. flavicaudus* may follow further gonadal differentiation in one of the two pathways: (a) Toward mature testis through degeneration of oocytes and development of spermatogenic tissue or (b) Toward mature ovary through growth of oocytes—very similar to the testicular and ovarian differentiation in the presumptive males and females of zebrafish. 2. (a) With the scarcity or absence of mature males in northern Vaipahu population, most fish with mixed stage of gonads postpone functioning as pure male, i.e., a sort of Okinawan hermaphrodism is maintained in which male-active hermaphrodites function as males and (b) With female's maturity at the smaller sizes of 3.2 cm in northern Vaipahu (as against > 5.2 cm in the southern Vaipahu and outer-reef populations), *D. flavicaudus* may prove to be a functional protogynic hermaphrodite. Asoh (2004) has also not ruled out the occurrence of sex change from vitellogenic oocytes to testis. However, it must also be noted that the evolution of sequential hermaphroditism from *Dascyllus*-like gonochorism is not a one-way traffic; there are also instances of hermaphroditic sparid *Leptoscarus vaigiensis* returning to gonochorism (see Pandian, 2010, 2011).

3.4 Structural and temporal costs

Simultaneous hermaphroditism is a rarity among vertebrates. It occurs only among 14 species (i.e., 0.05% species of fishes). It has to bear the costs of investments in morphological apparatus and functional mechanism of both sexes; competitive superiority of pure males could easily render hermaphroditism evolutionarily unstable (see Pandian, 2010).

Sex change among hermaphroditic fish is also a relatively rare phenomenon, and occurs only in 4% teleost families, i.e., 19 out of about 450 teleost families and includes about 109 (0.4%) species (Table 25). In these teleosts, sex change involves gonad, gonadal ducts and in some cases, secondary sexual characters (e.g., second ventral fin ray of *Cirrhilabrus temmincki;* anal fin color, *Pseudolabrus sieboldi,* see Pandian, 2010). Sex change is too costly that an α female changes to male only in the absence of physical presence and interactions of a territorial male. Interestingly, Lorenzi et al. (2006) exposed a harem of females of goby *Lythryprus dalli* to the (i) physical absence of the male, (ii) presence of the male behind a barrier that allowed visual and olfactory interactions but prevented physical interactions and (iii) presence of the male behind a porous screen that allowed olfactory interactions but prevented physical and visual interactions. The α-female changed sex only in the physical absence of the male. However, the effects of sex change alone are unlikely to explain the rarity of sex changers (Kazancioglu and Alonzo, 2009). Many theoretical considerations have regarded sex change as adaptive in the context of size-advantage hyphothesis.

Sex changers maintain sex ratios ranging from one male for every two or three females and thereby increase reproductive output. Analyzing male sex ratio among the different patterns of protogynous hermaphrodites, Pandian (2010) found reductions in male ratios to 0.31, 0.30, 0.23 and 0.22 in monoandric monochromatic, monoandric dichromatic, diandric dichromatic and haremic protogynous hermaphrodites, respectively. Quite independently, Molloy et al. (2007) too estimated the male ratio as 0.29 for 14 protogynous hermaphrodites. They have also estimated that the sex changing males invested just 20% of that of gonochores, while the females about 50% of that of the gonochores. Thus the protogynous sex changers seem to invest less on gonads, perhaps to neutralize the cost involved in sex change. The facts that theoretical considerations have not taken into account are (i) the sexually non-performing *transitional load* on the sexually active population and (ii) the fraction of life span, during which the transitionals persist in the population of protogynous hermaphrodites. Considering 16 protogynous species belonging to seven families, Pandian (2010) estimated the mean *transitional load* as 10.4% of the population and *transitional life span* as 31.6% at the mean male ratio of 0.33 (Fig. 42). In fact the male ratio seems to regulate the *transitional load* and percentage of *transitional life span.*

Various models have attempted to explain the curious but unique patterns of hermaphroditism in fish. The "low density model" explains simultaneous hermaphrodism as advantageous, despite the fact that they have to meet the high fixed costs for each sexual function and inherent antagonisms between male and female hormonal or other physiological system. The 'risk of movement model' has attempted to explain the existence of serial hermaphroditism; "intense predation pressure on patch-structured reefs make mate search movements very risky for small and sparsely distributed fish like gobies, thus giving a selective advantage to any stay-at home individual, who facultatively switch gender as the need arises" (Avise and Mank, 2009).

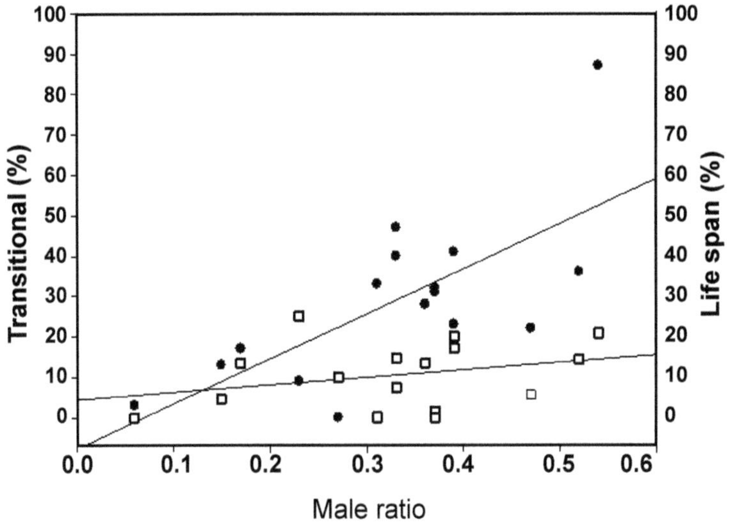

Fig. 42. Relationships between male ratio and transitional load (□) carried during the life spawn (•) of some protogynous hermaphrodites. (data are drawn from Table 20 of Pandian, 2010).

In formalizing the 'size advantage model', Warner (1975) has shown that if age-specific reproductive output increases more rapidly with body weight for one sex than the other and if the curves relating to fecundity with body size/age cross each other, then in principle an age or body size, at which an individual may reproductively profit by switching gender. Many scientists (e.g., Avise and Mank, 2009) have not recognized that body size-fecundity relation can differ from that of age-fecundity, and fishes may also experience menopause stage. Whereas fecundity increases with body size of fish, it does not during the period of terminal age (see Pandian, 2010, Fig. 43). When a fish has attained > 70% of its maximum age, it is approaching reproductively more or less inactive menopause stage at its terminal life span . In different populations of the guppy *P. reticulata*, this period of terminal life span has been estimated to range between 12 and 15% of total life span (Reznick et al., 2006). The empirical estimation predicts that sequential hermaphrodites change sex, when they reach 80% of their body size (Allsop and West, 2003). When all the recent information is taken together, it seems that the sequentials reach a body weight of 80% of their maximum, corresponding to the completion of 80–85% of their life span, when they produce negligible number of eggs (in relation to body weight and time, Fig. 27, 31) or cease to produce eggs, the protogynous hermaphrodites switch to males.

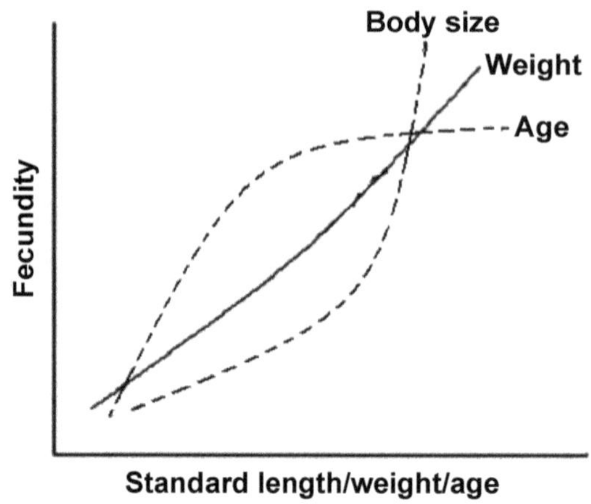

Fig. 43. Fecundity model, showing the relationships of fecundity to body size (-----), weight (—)and age (.....) of fishes (from Pandian, 2010)

3.5 Sex chromosomes and genes

Very surprisingly, cytogenetic studies on sex chromosomes on hermaphroditic fishes are limited to a dozen species only, though 380 or so hermaphroditic fishes including those in which functional hermaphroditism remains to be confirmed, are known to exist and their phenomenal diversity has allured considerable curiosity. Even among the dozen or so available publications, none has employed molecular cytogenetic studies like those of Ross and Peichel (2008) and Bradley et al. (2011). The conventional karyotypic studies thus far undertaken have only reported the number of chromosomes present (e.g., 2 = 48 *Sparus aurata,* Alvarez et al. (1991), or the lack of difference in the chromosome morphology during different sexual phases (e.g., monandric grouper *E. guttatus,* diandric wrasse *Thalossoma bifasciatum,* Ruis-Carus, 2002) or at the best the presence of 46 homomorphic chromosomes in *K. marmoratus* (Sola et al., 1997). Nevertheless, there are a couple of observations reporting the existence of female heterogametic serranid *E. tauvina* (2n = 48) and male heterogamety in the polymorphic labrid *C. julis* (2n = 46, 48, 46/45 XX/XO, see Devlin and Nagahama, 2002), although these observations require confirmation. Interestingly, the male heterogametic sex chromosome system of *C. julis* is XX/X0. It also remains to be tested whether the haremic male hetrogametic *C. julis* is a somatic female, like the nematode *C. elegans,* in which due to non-disjunction of a X chromosome, males are generated (Zarkower, 2006).

Among the gonochores, the role of sex determination by the *Dmy/ Dmrt1bY* has been unquestionably confirmed in a couple of medaka fishes. The claim of *amh* in tilapia (Shirak et al., 2006), *Dmrt1* in sturgeon (Hale et al., 2010) and *Dmrt1/cyp21a2* in zebrafish (Bradley et al., 2011) as candidate genes for sex determination remains to be confirmed. So far no sex determination gene has been identified in any of the hermaphroditic fish.

By pedigree based genetic parameter estimates, Batargias (1998) has established a genetic correlation between body weight and sex change in the gilthead seabream *S. aurata*. Through an interesting investigation, Loukovitis et al. (2011) have brought evidence for the genetic control of sex change in this non-model seabream. Combining body growth and sex change to non-monogenic traits, they have demonstrated a fast and effective strategy for QTL (Quantitative Trait Loci) detection through comparative genomics and linkage analysis. A BAC sequence information for SNP detection to enable fine mapping of QTL has also been developed. Analysis of 10 paternal half sib-families comprising of 409 progenies has revealed the presence of two sex linked loci in saLG21 group, namely one QTL for body weight and another affecting sex determination.

By estimating sex specific growth from a comparative study of growth rings in otoliths of α females with highest potential for sex change in harems of different densities of the wrasse *H. miniatus*, McCormick et al. (2010) have noted that the α female underwent growth acceleration at sex change. Hence, it is likely that *S. aurata* change sex at specific body size, whereas in the haremic *H. miniatus* the α female undergoes growth acceleration at sex change.

3.6 Sex differentiation genes

The subfamily Epinephlinae, commonly known as groupers, includes 159 species in 15 genera; the genus *Epinephelus* alone comprises of 98 species. Groupers are widely distributed throughout the tropical and subtropical waters of the world. Being a favorite marine food-fish, they are commercially important. They are amenable for culture, grow fast and resist diseases. Hence there is a growing interest for aquaculture of groupers. Being protogynous hermaphrodites, the males are however not readily available. Chinese scientists seem to have launched a major program to understand the molecular mechanism of sex change in the groupers. Not surprisingly, almost all available information on *sox* genes concerns the orange-spotted grouper *E. coioides* (Table 26). Using cDNA plasmid libraries from pituitary, hypothalamus and gonads at different phases of the orange spotted grouper, many scientists have constructed a set of SMART (switching mechanism at 5′ end of the RNA transcript). To date, Chinese scientists have cloned and identified the following genes: 1. *sox3*, which plays a regulatory role

Table 26. Summary of *sox* gene complex in downstream sex differentiation in orange spotted grouper *Epinephelus coioides*

Remarks	Reference
Ecsox3 full length cDNA contains 1,152 bp with an open reading frame (ORF) of 900 bp and encodes a peptide of 300 amino acids. Its protein is detected in differentiating PGCs, oogonia and oocytes, and subsequently in differentiating PGCs, Sertoli cells of testis, spermatogonia A type, which includes stem cells	Yao et al. (2003, 2007)
sox9 protein is expressed only in Sertoli cells and is mainly localized in cytoplasam	Xia et al. (2007)
sox9 mRNA consists of 3,277 bp with a 328 bp 5′ and translated region and 1,511 bp 3′ untranslated region. The 1,437 bp (ORF) encodes a 479 amino acid protein; a 70 amino acid HMG box is identical to many gonochores, zebrafish and medaka. It is expressed in testis, ovary, brain and other tissues but its expression becomes stronger with 17αMT treatment	Luo et al. (2010)
sox11a from ovary and *sox11b* from brain have been isolated and cloned	Zhang et al. (2010)

in germ cell differentiation and oogenesis (Yao et al., 2007), 2. *Dmrt*, which stimulates spermatogenesis (Xia et al., 2007), 3. *p450arom A* and *B*, which are essential components of androgen-estrogen enzyme complex (Zhang et al., 2007), 4. *FSHβ*, which shows sexually dimorphic expression in ovaries and the transitional gonads/testis (Wang et al., 2004), 5. *GTHα*, a key pituitary hormone that regulates gametogenesis and gonad steroidogenesis, 6. *TSHβ*, a hormone gene involved in growth, metabolism and reproduction (Wang et al., 2004) and 7. *PACAP*, which induces growth hormone and gonadotropin release from the pituitary (Jiang et al., 2003). Of these, (i) thyrotropin, a thyroid stimulating hormone β subunit (*TSHβ*), (ii) *EcDmrt1* and (iii) *sox3* are immediately relevant to sex determination and differentiation (Zhou and Gui, 2008). Figure 44 is a tentative illustration of the genetic cascade through which sex is determined in *E. coioides*.

(i) *sox*: The presence of *sox3, sox9, sox11a* and *b* has been reported from *E. coioides* (Table 26). These *sox* genes are similar in sequences but the deduced products vary in polypeptide length. They are expressed both in the ovary and testis. Obviously, the *sox* genes are involved in regulation of gametogenesis. The amino acid sequence of *sox3* shows 90% identity to *sox3* of medaka (Koster et al., 2000) and contains a highly conserved HMG domain of 79 amino acid in its N terminal region. Its HMG-domain shows 100% identity to that of medaka (Koster et al., 2000). The 70 amino acid HMG box of *sox9* is also identical to that of many fishes. Phylogenetic analysis based on full length of sox9 protein of many species revealed that *sox9* of fishes can be grouped into two clusters: cluster 1 includes gonochores

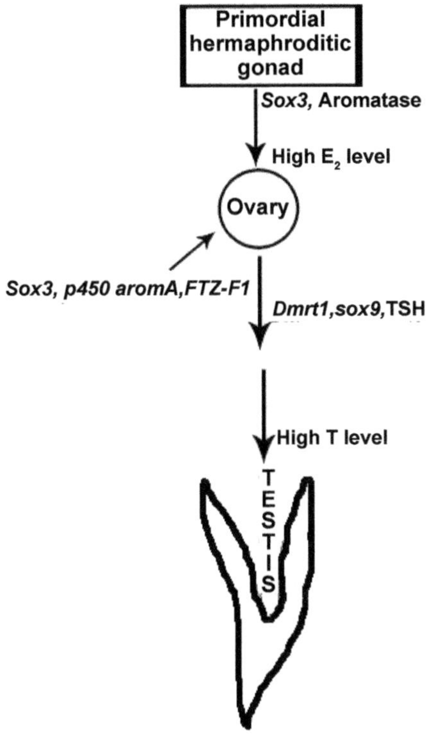

Fig. 44. A hypothetic model of genetic and endocrine mechanisms of protogynous sex change in hermaphroditic groupers. ← = inhibition, → = activation

like *G. aculeatus sox9b, O. latipes sox9b* and *D. rerio sox 9b* as well as protogynous hermaphrodites *E. coioides sox9* and *M. albus sox9a1*. The cluster 2 includes *D. rerio sox9a, G. aculeatus sox9a, O. latipes sox9a* and *Monopterus albus sox9a* (Luo et al., 2010). Clearly, both gonochoristic and hermaphroditic fishes employ the same sex differentiation genes. *sox9* is a sex related transcription factor and the earliest known marker of Sertoli cell differentiation. Over-expression of *sox9* causes male sexual development in an XX individual (Gardner et al., 2003).

In *E. coioides sox3* expression is detected very early in the nuclei of some PGCs along germinal epithelium. *Ecsox3*-positive PGCs develop toward spermatogonia. But on cessation of *Ecsox3* expression, the expressed *Ecsox3* are concentrated in one or two signal spots and are located outside the nuclei. *Ecsox3* transcripts begin to express at the lowest, as detected from MT-treated sex reversing grouper. The interesting data of Yao et al. (2007) indicate that as the transcription factor, *Ecsox3* has a more important role in oogenesis than spermatogenesis. *sox5* is involved in expressing Ikappaβ, a protein that is highly expressed in developing male gonad (see Gardner et al., 2003).

(ii) *Dmrt*: Table 27 shows that *Dmrt1* expression pattern is intimately linked to male gonadogenesis and further differentiation in both protogynous and protandrous hermaphroditic fishes. In *M. albus* too its expression is stronger in the testis than in the ovotestis and ovary. *Dmrt1* is not only differentially expressed and up-regulated in gonads of different phases in female to male sex change, but also it is restricted to specific phases and specific cells of spermatocytes stage

Table 27. Summary of *Dmrt1* gene complex in downstream sex differentiation in sequential hermaphroditic fishes

Species/Reference	Gonadal expression	Expression levels	Expression location
	Protogynous hermaphrodites		
Monopterus albus Huang et al. (2005)	Testis, Ovotestis ovary	High in testis	Differentiating testis
Epinephelus coioides Wang et al. (2004)	Testis	Higher in testis	-
E. coioides Xia et al. (2007)	Testis	-	Spermatogonia, Spermatocytes
Halichoeres tenuispinis Jeong et al. (2009)	Testis	-	-
	Protandrous hermaphrodites		
Acanthopagrus schlegeli He et al. (2003)	Testis	Higher in mature testis	-
Sparus auratus Liarte et al. (2007)	Testis	Decrease during sex change	-

I and II; however, it is not expressed in the supporting Sertoli cells. At the initial phase of sex reversing *E. coioides*, no Dmrt1 protein is detected; with continued MT-feeding and consequent sex reversing, *Dmrt1* begins to express and expression progressively increases during the course of 5 weeks of MT-feeding. Significantly, the protein has been detected only in spermatogonia and spermatocytes of the sex reversed male but not in its spermatids, an expression profile very much similar to that in zebrafish (Guo et al., 2005).

Both phylogenetic analysis and substitution estimates have revealed that *EcDmrt1* is related to that of the protandrous *A. schlegeli* (He et al., 2003) protogynous *H. tenuispinis* (Choi et al., 2004) and *M. albus* (Huang et al., 2005) and temperature sensitive *O. niloticus* and *O. bonariensis* (Miranda et al., 2001). No intron has been found in the ORF of *EcDmrt1* (Xia et al., 2007). Rhee et al. (2008, 2011) investigated the expression patterns of many sex differentiation related genes in the selfing hermaphrodite *K. marmoratus*. On exposure to E_2, the testis-specific *Dmrt1* mRNA has been found to be down-regulated in the gonad of secondary male but *dax1* mRNA has been significantly up-regulated. In the protandric *Sparus aurata,* the Dmrt1 mRNA levels are very low during testicular involution (Liarte et al., 2007).

(iii) *amh*: During gonadal differentiation of *A. schlegeli*, Wu et al. (2010) have detected the expression of *amh* and *amh2*, the expression level being higher in testicular tissue than in the ovarian tissue. There was no difference in the levels of expression of *amh* and *amh2* during active and inactive ovarian tissues. While the transcripts of *amh* are expressed in somatic cells, spermatogonia and vitellogenic oocytes, the expression of *amh2* is restricted to somatic cells of spermatogonia alone. The testicular expression of *amh* remains high, when the fish remains a male. Clearly, the *amh* is involved in regulation of gametogenesis during both male and female phases of the porgy but *amh2* is more intimately involved in spermatogenesis.

(iv) *FTZ-F1*: Two *FTZ-F1* are homologues of mammalian *NR5A1* and *NR5A2*. In response to MT treatment for 40–60 days, the expression of *FTZ-F1* decreases in the gonad but not in the pituitary. When tested *in vitro* on ovarian tissues, MT does not affect the expression of either *FTZ-F1* or *p450 aromA*, suggesting that the inhibition of gonadal p450 aromatase and *FTZ-F1* may be mediated at the brain-pituitary-gonadal axis (Zhang et al., 2007).

(iv) *TSHβ* : Its role in sex differentiation of *E. coioides* has been revealed for the first time but remains to be detected in other hermaphrodites. *EcTSHβ* mRNA is abundantly expressed and its expression level is higher in the testis than in the ovary.

(v) *cyp*: *cyp19a1a* and endogenous estrogen play a critical role in natural sex change. Progressively increasing expression levels of *cyp19a1a* have been reported from many protandrous species (e.g., *Sparus aurata,* Wong et al., 2006) and a reciprocal trend is observed in protogynous species (e.g., *E. akaara,* Li et al., 2006). The publications by Lee et al. (2008) and Wu et al. (2008) have provided a detailed account on the dynamic changes in the levels of *cyp19a1a* transcription and protein in the male, intersex and female black porgy. The dynamic changes in the levels from an undifferentiated gonad → differentiated testis → transforming bisexual gonad and finally to functional ovary suggest that *cyp19a1a* plays an important role in sex change from male to female in the porgy. Lee et al. (2006) have shown that the expression of *rmcyp19a* and *rmcyp19b* genes is differently modulated according to estrogenic compounds and gender type of *K. marmoratus.*

Liu et al. (2009) have cloned the key steroidogenic genes *p450 arom* and *p45011β* of *M. albus.* These genes show high homology to those of other fishes. The phylogenetic tree of *p45011β* cluster with the *p45011β* clades of medaka and rainbow trout. The expression level of *p450 arom* is 17 times higher in the ovary than in the testis but that of *p45011β* is faint in the ovary, slight in the ovotestis and abundant in the testis. With concomitant changes in the expression of *p450 arom* and *p45011β*, sex of the eel is changed from female to male.

Recently, a clear sexually dimorphic expression of *p450 aromatase (cyp19a1a)* has been described in the yellow tail clownfish *A. clarkii* by Kobayashi Y, et al. (2010a). The full length cDNA of the *p450 aromatase* gene spans to 1,928 bp long and encodes 520 amino acids. No signal of *cyp19a1a* is detectable in the bisexual gonad. But a high level of *cyp19a1a* transcripts (about 2,300 copies/ng total RNA) is detected in the ovary of the female phase. *cyp19a1a* immunoreactivity is confined exclusively to the granulosa cells. A phylogenetic tree depicting the relationship of various cyp19 proteins shows that the clownfish falls into the cluster of grouper, wrasse, sea bass and flounder.

(vi) *foxl2*: Kobayashi et al. (2010c) have cloned the *foxl2* cDNA of the protogynous wrasse *H. trimaculatus*. The isolated *foxl2* cDNA is 1989 bp long, encodes 306 amino acid proteins and shows 90% homology to that of fishes belonging to the order Perciformes. Figure 45 reveals that the Forkhead (FH) domain, also known as 'winged helix', is highly conserved among vertebrates. Using zebrafish *foxl1* as outgroup, the phylogenetic tree constructed shows that vertebrate *foxl2* sequence, except for rainbow trout, shares high bootstrap values. *foxl2* is known to specifically express in the ovaries. As expected, the wrasse *foxl2* mRNA is expressed in the ovary during the female phase and signals of foxl2 protein are localized in granulosa cells but not in thecal cells; *foxl2* is shown to regulate *p450 arom* expression in the ovary. However, the *foxl2* mRNA transcript levels in the testes of IP and TP males are almost equal to that of the ovary suggesting the role of *foxl2* may be quite different from other gonochoric vertebrates; in the sex changing fishes, *foxl2* controls the steroidogenesis.

(vii) *sf-1/nr5a1* and *nrOb1/Dax-1*are important genes involved in steroidogonesis and sexual differentiation. In *A. schlegeli, nr5a1* but not *nrOb1* is required for differentiation of the early testis. High expression levels (3–11 times of the initial) of *nr5a1* favor the maintenance testicular differentiation during the male phase. In contrast, low *nr5a1* and high *nrOb1* expressions sustain sex change to the female phase. Obviously, *nr5a1* and *nrOb1* have a cooperative function for testicular differentiation and subsequent antagonistic interaction for oocyte development (Wu et al., 2008).

To understand the underlying mechanism of hypothalmic-pituitary-gonadal axis involved in gonadal sex change in the honey comb grouper

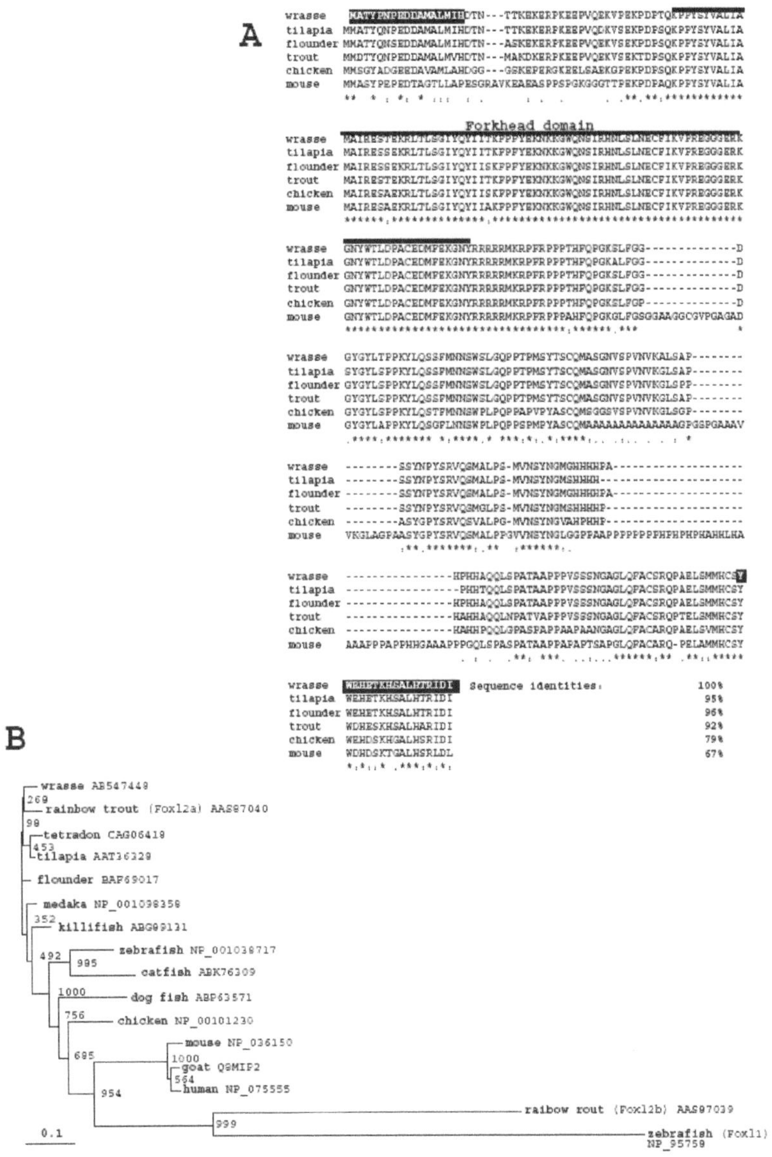

Fig. 45. Sequence and phylogenetic analysis of *foxl2* proteins. (A) Sequence alignment of *foxl2* proteins. Forkhead domain is underlined. The identification of wrasse *foxl2* sequence with other sequences is indicated in percentage against each sequence. Amino acids identical to those in other species are indicated by asterisks (*). Colon (:) indicates conserved substitution and dot (.) indicates semi-conserved substitution. (B) A phylogenetic tree comprising the subfamily of *foxl2* proteins. GenBank Accession Numbers for the aligned proteins are given with species name. Horizontal line indicates genetic distance (from Kobayashi Y et al., 2010c)

E. merra, Kobayashi et al. (2010b) have studied the expression pattern of mRNAs for gonadotropin (GTH) subunits *Cga, fshb and fhb* in the pituitary at different sexual phases. The *fshb* mRNA level, which remains low during the female phase, dramatically increases with concomitant development of the testis. After 3 weeks, FSH treatments (50–500 mg/g fish) induce the female to the male phase and up-regulate endogenous androgen levels and *fshb* transcripts, while LH treatment shows no effect. Hence FSH may trigger the female to male sex change in the groupers (see also Alam et al., 2010). It remains to be seen whether LH triggers the male to female sex change in protandrics.

Sex steroid receptor genes: In the protogynous *Halichoeres trimaculatus, ar mRNA* levels are higher in the testis but decrease in the following order: functional ovary < bisexual-gonad and its ovarian tissues < testis (Kim et al., 2002). But its level in the protandrous black porgy *A. schlegeli* is higher prior to sex change (He et al., 2003). Taken together, it is clear that the androgen receptor is important for the male phase in both protogynous and protandrous hermaphrodites.

Estrogen receptor cDNAα type has been cloned from the ovary and liver of *A. schlegeli* (Lee et al., 2001). The *erα* is significantly more strongly expressed in the vitellogenic ovary than in that of immature ovarian tissue. There is no difference in the level of *erα* in the functional male and bisexual gonad. The *er* is also expressed testicular gonad of the porgy and is consistent with that reported for *O. mykiss* (Baroiller et al., 1999) and a combination of *erα* expression and at low doses (0.25 mg) of E_2 significantly increases the number of spermiating males, milt volume and GSI of the porgy (Chang et al., 1995).

In view of the role of Hypothalamic-Pituitary-Gonad (HPG) axis in sexual maturation of fishes, it is considered that the HPG axis may also be involved in sex change among the hermaphroditic fishes. For example pre-optic GnRH and Arginine Vasotocin (AVT) axis have been implicated as responsible for sexual plasticity in fishes (Foran and Bass, 1999), monoamine transmitters as a stimulating factor in sex reversal of *T. duperrey* (Larson et al., 2003), neuropeptide Y (NPY) as an inducer of sex reversal in *T. bifasciatum* (Kramer and Imbriano, 1997) and hGC as an effector of sex reversal in *Coris julis* (Reinboth and Brusle-Sigard, 1997). The gobiid fish *Trimma okinawae* is known to rapidly change its sex back and forth from male to female and then to male serially, depending on the social status in the harem. Kobayashi et al. (2009) investigated the role of gonadotropin receptor (*GtHR*) gene during the onset of sex change from female to male and male to female. They also examined the role of *FSHR* (Follicular Stimulating Hormone Receptor) and *LHR* (Luteinizing Hormone Receptor) genes, as the action of GtH is mediated through its receptors *FSHR* and *LHR* located in the gonad. The gonad of the goby has both ovaries and testes simultaneously. During the male phase, the testicular portion is fully differentiated and contains mature sperm, and the ovary is in a rudimentary state. The opposite scenario is observed during the female phase. Levels of goby *FSHR* and *LHR* transcripts have been determined as a measure of gene expression in the ovary and testis during sex change and are expressed as ratio of *FSHR* and *LHR* to *GAPDH* gene. Expectedly both *FSHR* and *LHR* are highly expressed only in the gonad corresponding to the sexual phase in a typical haremic goby.

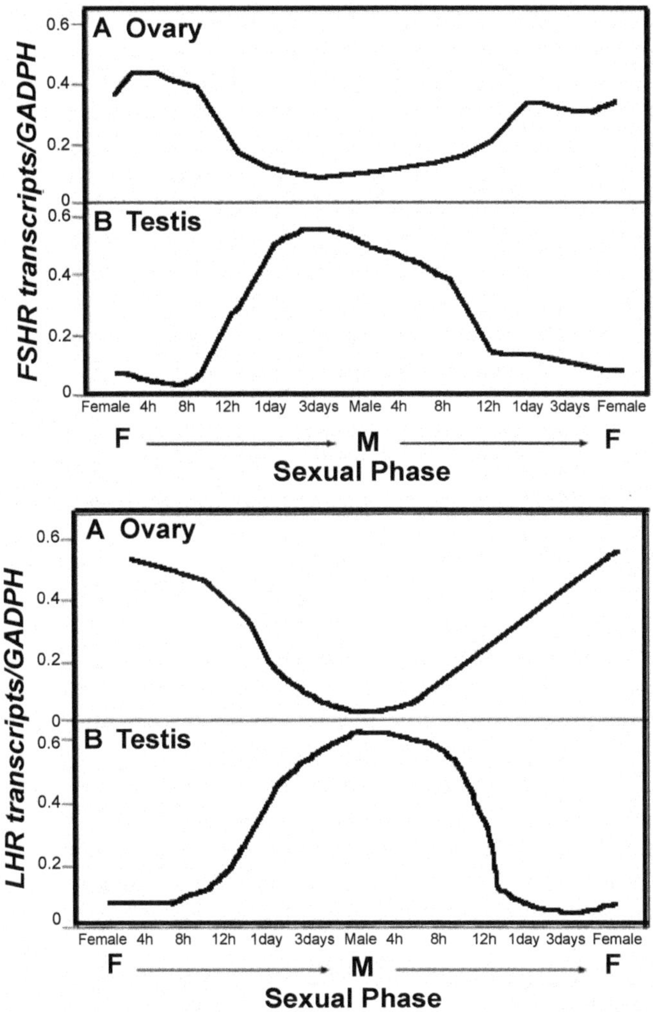

Fig. 46. Schematic illustration of the trends observed for (upper panel) FSHR and (lower panel) LHR as function of sexual phase during sex change in a common haremic goby (compiled from Kobayashi et al., 2009)

Figure 46 shows the chronological sequence of changes in *FSHR* and *LHR* levels in the ovary and testis, when sex is rapidly changed from female to male and then to female each within a short span of four days. These findings suggest the important role played by the hypothalamic-pitutary-gonadotropic axis in realizing rapid sex change.

3.7 Gonadal differentiation

The occurrence of hermaphroditism in fishes is well documented (e.g., Pandian, 2010) and are common among, for example, Sparidae, Scaridae and Serranidae, in which hermaphroditism is the rule rather than exception (Buxton, 1989). The hermaphrodites may have arisen independently in more than 10 lineages (Warner, 1978). In accordance with different patterns of hermaphroditism, a brief account on gonadal differentiation is provided.

Self fertilizing hermaphrodite: *Genetically different groups* (Turner et al., 1992) of *homozygous isogenic clones* of *Kryptolebias marmoratus* naturally reproduce *uniparentally* in waters from southern Florida through much of the West Indies to Venezuela (see Pandian, 2010). These attributes are shared by no other vertebrates known so far. Following self fertilization (Fig. 47), the eggs are incubated intraparentally for 2–3 days and then extraparentally for 14 days (Harrington, 1967). On incubation at 26°C, hermaphrodites are produced but primary male, when incubated at < 19°C. Exposure of hatchlings to 28°C generate secondary males. In its natural habitats, secondary males may fertilize 2–3% hermaphrodites (Tatarenkov et al., 2009).

The estimates of Germ Cells (GCs) in rivulus is perhaps the only contribution to date reporting on dimorphic trends of the mitotic proliferation of GCs as a function of early embryonic and post-embryonic stages in hermaphrodites and males. Harrington (1975) considered body length as more appropriate than time scale, as the latter is subjected to wider changes, especially due to differences in incubation temperature. Figure 48, reconstructed and redrawn using Harrington's data, clearly shows the sexually dimorphic trends for the mitotic proliferation of GCs (approximately the PGCs) in hermaphrodites and male rivulus. Though accelerated from 30–31 stages, the proliferation of GCs rapidly increases to about 680 in hermaphrodites but to about 340 only in males. These trends unequivocally resemble those of PGCs in *O. latipes* and *O. niloticus* (see Fig. 13). Hence it is likely that sex in the selfing *K. marmoratus* is also determined by *Dmrt*. A publication confirming this suggestion would make a milestone in this area of research. Indeed Rhee et al. (2008) have reported the presence of *Dmrt1* in the rivulus.

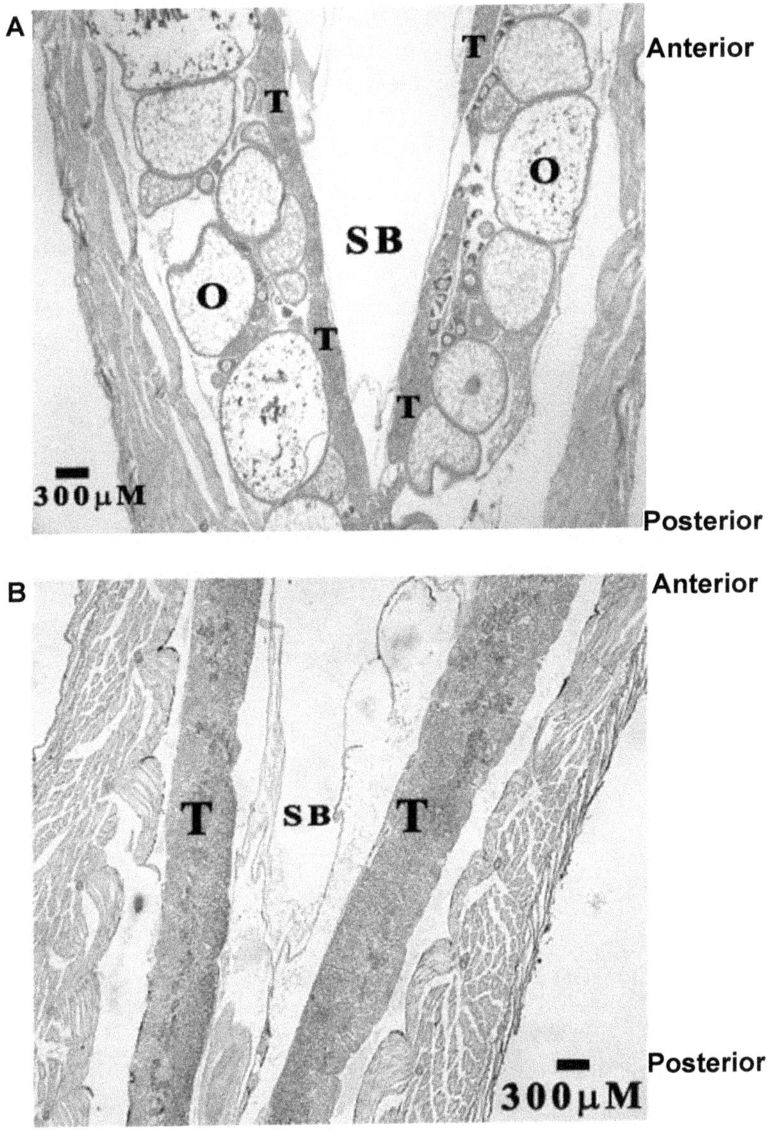

Fig. 47. Ventral sections of (A) hermaphrodite and (B) male rivulus *Kryptolebias marmoratus*. T = testicular tissue, O = ovarian tissue, and (SB) swim bladder (from Orlando et al 2006. Cloning and differential expression of estrogen receptor and aromatase genes in the self fertilizing hermaphrodite and male mangrove rivulus *Karyptolebias marmoratus*. Journal of Molecular Endocrinology, 37: 353–365, reproduced by kind permission of Society for Endocronology, 2011)

At 26°C, the rivulus hermaphrodites are sexualized as males, (cf Sakakura and Noakes, 2000) at the size of > 5.3 mm or older than 36 days. But the males attain sexual maturity only at the size of > 6.5 mm and age older than 76 days. The smallest male at sexual maturation is 2 mm longer and 48 days older than the youngest and smallest hermaphrodite. However, the hermaphrodites and primary males attain functional maturity at 122 and 206 dph, respectively. The juvenile rivulus have a caudal ocellatus, persistent in adult hermaphrodite but obsolescent or lacking in both primary and secondary males. Hermaphrodites have prominent body markings and barred fins. These are retained in secondary males, which acquire an additional orange spot. The primary males are bright orange in color. Figure 48 shows the structural organization of gonads in hermaphroditic rivulus.

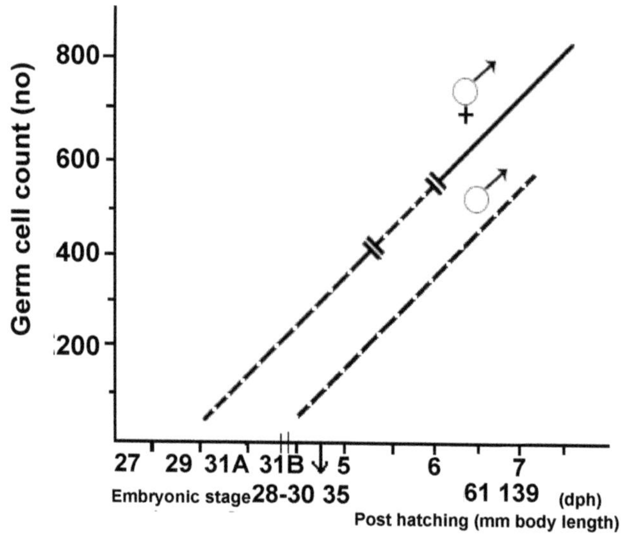

Fig. 48. Mitotic germ cell proliferation in embryos/hatchlings of *Kryptolebias marmoratus* incubated at 19°C and 26°C leading to differentiation of males (-------) and hermaphrodites (—), respectively. The numbers 27–31A, 31B represent embryonic stages, 30 dph a body size of 4.8 mm on 30 day post hatching (figure redrawn by compiling data of Harrington, 1975)

Recent studies indicate that immature rivulus (< 60 dph) are indeed females and the frequency of hermaphrodites increases with advancing age (Sakakura and Noakes, 2000). The sex change from female to hermaphrodite can be prevented by immersion of the juveniles in 1ppb ethynylestradiol during the first 28 days. Hence sex steroids, their receptors and aromatase play an important role in sex differentiation. The rivulus have *erα*, *erβ* and *aromA* and *aromB* genes that are more strongly expressed in the liver, brain and gonad of hermaphrodites than those in males (Orlando et al., 2006).

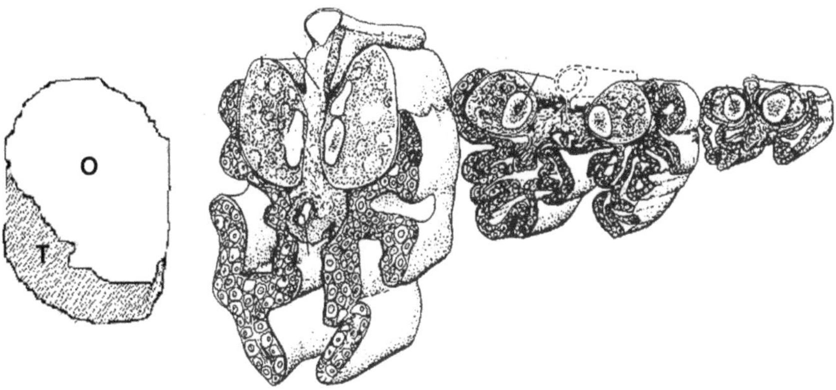

Fig. 49. Left panel: Cross section through the gonad of a serranid simultaneous hermaphrodite. Right panel: Bisexual gonad of *Siderea grisea* as schematically viewed from the left side. Right, the most posterior, left, the most anterior. OL- ovarian lamella; T- testis, SUS- vein; A- artery; ST- spermatogonial tissue; SD- 'sperm duct' (from Fishelson, 1992)

In gamete exchanging hermaphrodites, the ovarian and testicular tissues are located in heterologous zones (Fig. 49) with separate gonoducts for either sex opening to the exterior (Fig. 49); hence the chances for self- fertilization is eliminated. However, the presence of several spermatozoa on a hydrated oocyte within the ovarian lumen of *Serranus atricauda* suggests that the chances of internal fertilization cannot totally be excluded (Garcia-Diaz et al., 2002). Incidentally, earlier observations summarized by Atz (1964) are interesting and relevant. Individuals of *S. scriba* and *S. subligarius* kept in aquaria more or less simultaneously emitted eggs and milt, and the eggs were externally 'self-fertilized'. Eggs and sperm collected from the same individual of *S. scriba* were also amenable for 'self-fertilization'. However, cross-fertilized eggs survived better. Hence externally cross-fertilized eggs of these gamete exchanging hermaphrodites have better fitness than the possible internally or externally self-fertilized eggs.

Sex changers: Sequential hermaphroditism comprises of female to male sex changing protogynous, and male to female sex changing protandrous hermaphrodites (Fig. 2). The sex change results in secondary males in the former and secondary females in the latter. Among some of these sequentials, primary males and females, who do not pass through the female and male phases, respectively, are directly generated in protogynous diandrics (e.g., *T. bifasciatum*) and protoandrous digynics (e.g., *Lates calcarifer*). Among the monandric protogynics, monochromatics (e.g., *E. flavolimbatus*) and

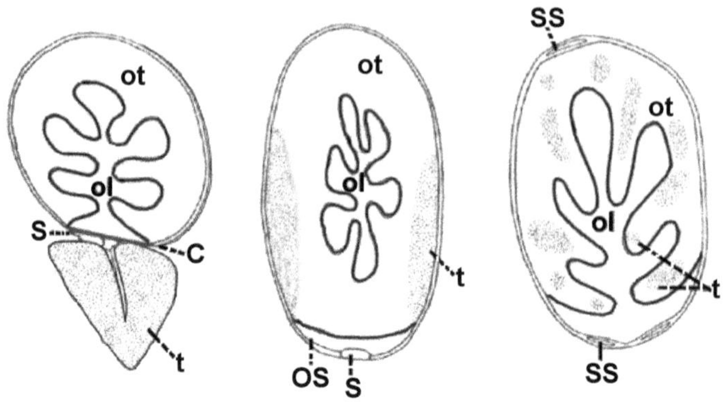

Fig. 50. Changes in the configuration of ovarian and testicular tissues in hermaphroditic fishes. (a) Delimited type: *Pagrus pagrus*, (b) Undelimited type 1: *Hypoplectrus puella* and (c) Undelimited type 2: *Epinephelus guttatus*. c = connective tissue; ol = ovarian lumen; os = ovarian sinus; ot = ovarian tissue; s = sperm duct; ss = sperm sinus; t = testicular tissue (from Sadovy and Shapiro, 1987)

dichromatics (e.g., *Sparisoma viride*) are present. Among the diandric species, the newly sex changed Initial Phase (IP) males have a dull body color like the females (cf Neff, 2003a) but they are then transformed into brightly colored Terminal Phase (TP) males (e.g., *Calotomus japonicus*). The serials change sex more than once in either direction. One or another pattern of hermaphroditism is not necessarily restricted to a taxonomic family or genus. For example, *T. cupido* is a monoandric but *T. bifasciatum* a diandric, *Pseudolabrus celidotus* a dichromatic monoandric but *P. sieboldi* a serial (Pandian, 2010). Within a single genus such as *Choerodon*, there are monoandric, as in *C. transversalis* and diandric, as in *C. albigena* (Nakazona and Kusen, 1991).

Distribution of gonad types: Hermaphrodites have either an ovotestis or a gonad. Delimited type of ovotestis consisting of distinct heterologous zones of testicular and ovarian lobes, which are separated by a connective wall (Fig. 50) is present among both protogynous species (e.g., many sparids, *P. pagrus*) and protandrous species (e.g., *Acanthopagrus schlegeli*). In these sparids, the ovary is located dorsally and the testis ventrally. The undelimited type of gonads, are classified into two groups (Sadovy and Shapiro, 1987). In the undelimited type 1, the ovarian and testicular tissues come in contact with each other but are never intermixed (e.g., *Hypoplectrus puella*) and in the type 2, the ovarian and testicular tissues are intermingled during the course of sex change (e.g., *E. guttatus*). Rather a very different gonad type, not described by Sadovy and Shapiro (1987), has been described by Takamoto et al. (2003); in the balastid *Sufflamen chrysopterus* the female to male sex change is accompanied by the diminishing ovarian component, and enlarging and maturation of the vestigial testies (Fig. 51).

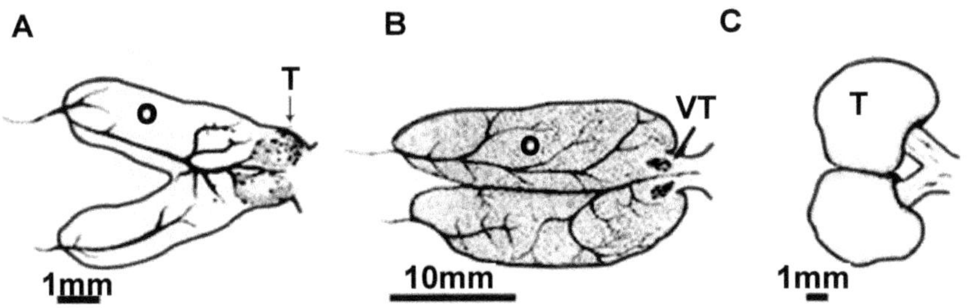

Fig. 51. *Safflamen chrysopterus*: Schematic representation of (A) gonads of a juvenile with female coloration, (B) a breeding female and (C) a breeding male O = ovarian part; VT = vestigial testicular tissue; T = testicular tissue (from Takamoto et al., 2003)

Considering the groupers as a model, and major events in ovarian differentiation, ovarian degeneration and testicular re-differentiation in protogynous species are briefly summarized in Table 28. In some sex changers, the transition in body color concomitant with sex change is spectacular; a large number of such astounding examples are illustrated by Lieske and Myers (1996), which may indicate the internal gonadal status of the transitional. For recent information, Reinboth and Brusle-Sigard (1997), Ohta et al. (2003) and Kline et al. (2011) may be referred. For more information on details of the ovarian transformation to testis, Pandian (2010) may be consulted. During the sex change, the testicular tissues develop from the former ovarian wall or proliferate from the wall bordering the ovarian

Table 28. Major events in ovarian differentiation in *Epinephelus coioides*, de-differentiation in *E. tauvina* and testicular re-differentiation in *E. fuscoguttatus* (compiled from Liu and Sadovy de Mitcheson, 2009; Mathew, 2011; Pears et al., 2007)

Ovarian differentiation	Ovarian de-differentiation	Testicular re-differentiation
Gonadal primordium, undifferentiated gonad	Ovary loose-walled, elongated and compressed	Testis elongated to a cleft or triangular, crypts of precocious spermatocytes and precursor testicular tissue appear
Ovarian phase 1–2 gonad, undifferentiated juvenile	Yellow brown bodies appear, emergence of granulocytes	Crypts of spermatozoa joined within testicular lobules to form central sperm sinuses
Ovarian phase 3 gonad, differentiating female	Atresia and degeneration of peri-vitelline and chromatin nuclear oocytes, gonad with intermingled spermatogonial tissue	Sperm ducts are formed, brown bodies and ovarian cavity persistent
Functional ovary	Oocytes reduced and filled with vacuoles, spermatogonia and spermatocytes present	Functional testis

cavity radially or linearly, as in gobiids (Cole and Shaprio, 1992). A unique feature of the reproductive system of male gobiids is the presence of paired secretory Accessory Gonadal Structure (AGS) associated with the testis. At the time of sex change, it arises from the precursive tissue masses (pAGS) associated with the ventral portion of the ovarian wall in the region of the common genital sinus. The ovarian cavity is not persistent in the testes of sex changed gobiids and parapercids (e.g., *Parapercis snyderi*, Kobayashi et al., 1993). Among the protoandrics, a thin layer of connective tissue separates the peripheral testicular and interior ovarian tissues in many protandrics like *Diplodus sargus* (Micale and Perdichizzi, 1994). However, the connective tissue is lacking in *A. frenatus* (Brusle-Sigard and Reinboth, 1990); male and female germ cells are separated by their respective somatic Sertoli and follicle cells only, and thereby pose challenging problems to endocrinologists. For information on the formation and development of gonadal ducts in the sex changers, Kobayashi et al. (1993), Lo Nostro and Guerrero (1996) and Alam and Nakamura (2007) may be consulted.

The ontogenetic pathways of sex change in the diandric protogynous sparid *P. pagrus* and a protandrus *A. schelegeli* are illustrated in Fig. 52. As can be seen, these two hermaphrodites have the delimited gonad type. In the presence of the delimited type of ovotestis and the consequent 'no conflict of hormones' the initial gametogenesis proceeds independently in the testicular and ovarian zones prior to sexual maturity. Following sexual maturity, a functional ovary alone is differentiated in the protogynics like *P. pagrus*, during the initial female phase; at the onset on sex change, the ovary is more or less completely regressed, prior to the testicular differentiation (Kokokiris et al., 2006). Contrastingly, the ovarian and testicular zones of the protandric *A. schlegeli* regress and re-differentiate tuned to the milting and resting seasons during the initial male phase lasting for 2–3 years (Lee et al., 2008). Surprisingly a similar pattern of alternate regression and re-differentiation of the testis and ovary is described for protandric *Diplodus sargus* possessing a gonad, in which "a thin layer of connective tissue alone separates the heterosexual zones (Micale and Perdichizzi, 1994) and possibly in *Sparus aurata* (Liarte et al. 2007)." Apparently, the processes of gonadal de-differentiation and re-differentiation occur only once following the completion of the initial female phase in the protogynics but more than once during the initial male phase itself in the protandrics (Fig. 53). The regressed testicular and ovarian zones are left as a small vestigial structure. Contrastingly, the gonad of secondary sex is simply derived from the existing wall of the primary gonad in almost all the diandrics (Fig. 54) and digynics. From the same gonad, which functioned as ovaries in protogynics and testes in protoandrics, the secondary testes and ovaries, respectively are derived. Available figures on derivation of testis from the former wall of the undelimited type of the ovary of many protogynic

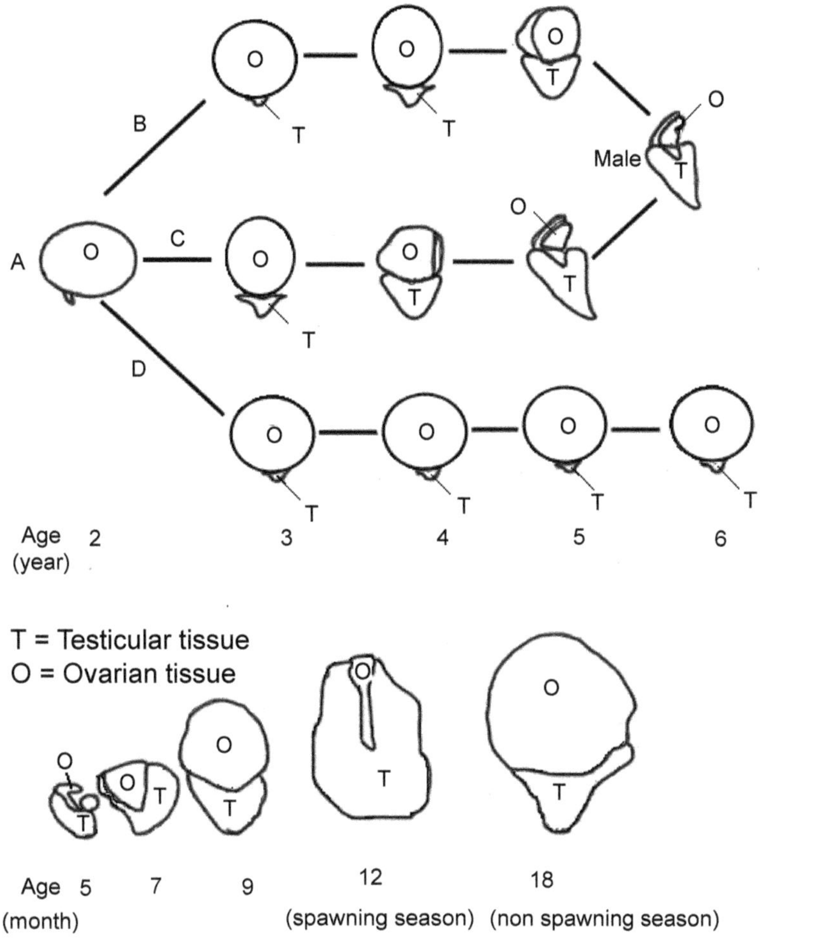

Fig. 52. Schematic representation of ontogenetic pathways of sex changes in hermaphroditic fishes. Upper panel: pathways of (A) egg (B) direct transformation to male through juvenile hermaphroditism, (C) indirect transformation to male through functional female and (D) retention of functional female status throughout the life span of protogynic (compiled from Kokokiris et al., 1999). Lower panel: Ontogenic pathway of transformation to a functional female in protandric (compiled from Chang et al., 1997)

species represent only a simple membrane (Fig. 54). The wall has not yet attracted electronmicroscopic studies (cf Lacerda et al., 2006, Fig. 41). It remains to be seen whether the former ovarian wall comprises of PGCs/SSCs and/or germ cells supporting somatic cells, from which cells of the spermatogonia arise. This information is a base for further experiments like the transplantation of PGCs/SSCs/OSCs.

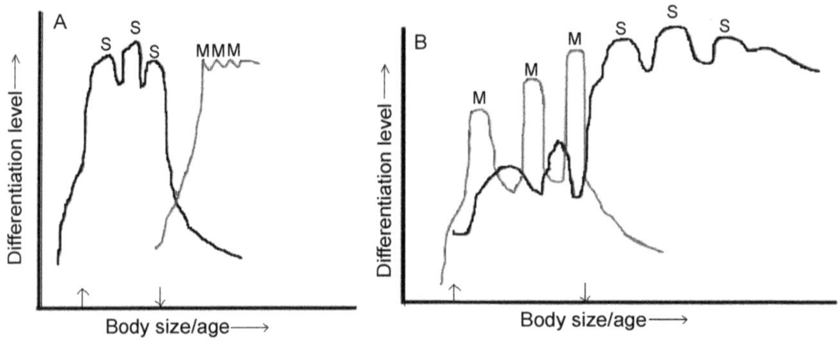

Fig. 53. Schematic illustration of structural sex differentiation levels as function of body size/age (A) protogynic and (B) protoandric hermaphrodites. S = Spawning, M = Milting, ↑ = size/age at sexual maturity, ↓ = size/age at sex change

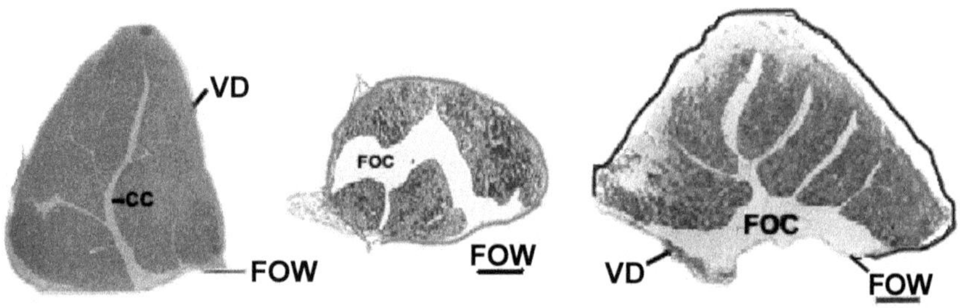

Fig. 54. Transverse section of testis of (A) *Lethrinus rubrioperculatus* (from Ebisawa, 1997) (B) *Calotomus japonicus* (from Kusen and Nakazono, 1991) and (C) *Thalassoma lutescens* (from Shibuno et al., 1994). Note the presence of former ovarian wall and former ovarian canal in these testes

Using nuclear and cytoplasmic diameters and their location in the germinal epithelium of the protandric *Sparus aurata*, Chaves-Pozo et al. (2005) distinguished Spermatogonial Stem Cells (SSCs) from Primary Spermatogonial Cells (PGCs). Both Sertoli Cells (SCs), SSCs and PSCs proliferate by successive spermatogenetic waves throughout the reproductive cycle at the reproductive stage-dependent rate. During post spawning, the resumption of proliferation, the occurrence of apoptotic spermatogonia, and the phagocytosis of nonshed spermatozoa by SCs lead to a reorganization of both the germinal compartment and the interstitial tissue. But most unexpectedly, both oogonia and oocytes are also proliferated, suggesting the operation of a complex mechanism in the regulation of gonocyte proliferation in protoandric fishes.

The foregone descriptions have shown that there is wide structural diversity in the gonads and possible gonadal ducts of sex changing hermaphrodites. Yet there seems a commonality between gonochores and hermaphrodites, and uniformity in the endocrine mechanism which realizes the expression of sex determination and differentiation genes and regulates the synthesis and action of the endocrines. Importantly, estrogens act as an ovarian inducer and play an important role in protogynous sex change (Nakamura et al., 2003). Aromatase is the critical enzyme that must be present for biosynthesis of E_2 from testosterone. In contrast, the lack of steroid hormones, including androgen, plays an important role in testicular differentiation. All the Nile tilapias treated with aromatase inhibitor (AI, e.g., tamoxifan, fadrozole) have hermaphroditic gonad with both ovarian and testicular tissues; clusters of spermatogonial cells distributed around the outer periphery of the ovigerous lamellae and clusters of Leydig cells are observed, indicating that "(i) it is possible to induce functional sex reversal of the gonads, prior to the commencement of sex differentiation, i.e. the labile period the completion of sex differentiation and (ii) both germ cells and germ cells supporting somatic cells have still retained the bisexual potency until the closure of labile period" (Nakamura et al., 2003).

3.8 Morphotypes and differentiation

Diandric and digynic species generate primary males and primary females, respectively, perhaps to compensate the progressively decreasing male ratio in protogynic species and female ratio in protandric species (Pandian, 2010). Sperm limitation is known to occur and reduce fertilization success among pair-spawning species (Shapiro et al., 1994); it has implications to the probability of egg-sperm encounters, especially where the egg size is increased resulting in the reduction of fecundity and duration of floatation due to the possible increase in egg density. A primary male may release 7.5 times more sperm than that of a secondary male (Shapiro and Rasotto, 1993) and motility of the sperm of the primary male is several times faster than that of secondary males (Fitzpatrick et al., 2007). Consequently, the fertilization success of pair-spawning diandrics such as the tropical wrasse *Halichoeres bivittatus* is 78%, against 85% in the presence of one or more streaking primary males (Peterson, 1991). Hence the presence of primary males during 'spawning rush' may ensure a higher fertilization success.

From detailed histological studies, Liu and Sadovy de Mitcheson (2009) have described major events of gonadal differentiation in monandric and diandric groupers. An undifferentiated juvenile with biphasic gonad may choose alternate pathways of sexual differentiation in a diandric grouper: differentiate directly into a primary male or a functional female (Fig. 55). Primary males resemble the females in body coloration and appearance but the secondary males, have a bright body color during the terminal phase (TP). Shapiro and Rasotta (1993) indicate the instances of primary males of *T. bifasciatum* changing their (intial phase, IP) color into TP males and the

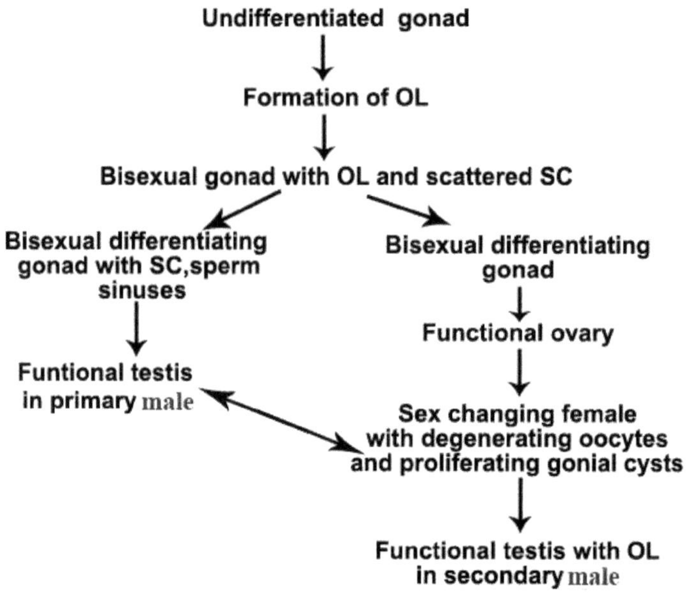

Fig. 55. Schematic illustration of gonadal development and differentiation in primary male and female in a diandric protogynic hermaphrodites. The gonadal differentiation in diandrics pass through an initial juvenile hermaphroditism. In others, primary males can change to female and a female also can switch over to primary male. OL = Ovarian lumen, SC = Spermatogonial cysts

females too re-differentiate into a 'secondary male'. Munday et al. (2006) have shown that the juveniles of *T. bifasciatum* may choose to differentiate into primary males or females, depending on the female population size in a location. From their long time field and aquarium observations, Kuwamura et al. (2007) have shown that a primary male of *H. trimaculatus* changes to female and an adult female, even after a few spawnings, changes into primary male, i.e., *H. trimaculatus* is a bidirectional serial hermaphrodite. Kojima et al. (2008) were the first to sex reverse the primary IP male to female of *H. trimaculatus* by E_2-treatment. Following the treatment, spermatogenic germ cells in the testis underwent rapid degeneration with concomitant appearance of female germ cells. The E_2-treatment suppresses the expression of genes involved in steroidogenesis in the testis such as 11β hydroxylase and *Dmrt1*, and reduces androgen receptor transcripts.

These observations have clearly revealed that (i) the morphotypes, such as primary male (IP), functional female and TP male are only phenotypes and are not genotypes, as has been considered for a long time and (ii) these morphotypes have retained the bipotential gonadal germ cells and germ cells supporting somatic cells, even after sexual maturation and miltings or spawnings: Most remarkably, these phenotypic morphotypes among hermaphrodites can change from primary males to females and females to primary IP males and secondary TP males. Hence these sex changes are radically different from those among gonochores, in which the phenotypic morphotypes change at the best from sneaker male to satellite male, satellite male to territorial male and/or territorial male to pirate male—all these changes occur within a single sex namely male. Thus the changes in morphotypes are *intrasexual* among gonochores but are *intersexual* among hermaphrodites. This generalization may implicate that the flexibility/plasticity of sex determination and differentiation processes are several times greater among hermaphrodites than among gonochores.

Incidentally, the gonadal differentiation in primary gonochoric fishes implies the loss of bisexual potency by the germ cells supporting somatic cells. On the other hand, the bisexual potency is retained by the somatic cells of the secondary gonochores and hermaphrodites, whose brains are not sexualized with the presence and activity of radial glial cells, as in zebrafish. Yet the hermaphrodites are not monophyletic and are suggested to have arisen independently in more than 10 lineages (Warner et al., 1978). Expectedly, the bisexual potency of the somatic cells in the few but widely diverse group of hermaphrodites, presumably also remains diverse. For instance, not all the females of *P. pagrus* (see Fig. 52) undergo natural sex change throughout their life time. The same holds true of the protandrics like *Diplodus vulgarus* (see Pandian, 2010). Does it mean that the bisexual potency of the somatic cells is lost in them? *P. pagrus* is amenable for rearing in laboratory (Kokokris et al., 1999). It is then for the French neuroendocronologists to investigate the presence and activity of radial glial cells of the red porgy; for it seems that the radial glial cells control the retention of bisexual potency by the germ cells supporting somatic cells.

4

Sterility and Mutants

The role played by gonads and their genes in the process of sex differentiation is realized in their absence. Gonadectomy is one of the earliest techniques used to know the role played by the gonads and their hormones in sex differentiation. Because of their rarity and inability to function, the abnormal hermaphrodites, though an important resource, have so far not received much attention. This is perhaps true of the steriles too. Morbid embryos, caused by one or another mutant gene, have long been reported. However, the mutants, especially the viable ones, have begun to receive their due attention by medical and molecular biologists since this millennium. This section broadly indicates their occurrence and importance in our understanding of the process of differentiation including sex differentiation.

4.1 Gonadectomy

Castration/ovariectomy is one of the earliest techniques employed to understand the role of hormones in sexualization of the gonad. Turner (1947 cited in Atz, 1964) has reported the complete regeneration of gonopodium in castrated *Gambusia affinis* with regenerated testis as small as 4% of the normal size. Clearly, the very presence of the masculinizing hormone like 11β-hydroxytestesterone (see Table 1) in the small regenerating testis is adequate for regeneration of the gonopodium. Impressed by the findings of Noble and Kumpf (1937 cited in Atz, 1964), the zealous Reinboth (cited in Atz, 1964) recorded the apparently accelerated appearance of testis in a unilaterally ovariectomized protogynic female *Coris julis* beginning to change sex. A better designed gonadectomy in *C. julis*, which is amenable to surgical gonadectomy, may prove to be rewarding.

Unilateral Gonadectomy (ULG) entails surgical removal of one gonad and leaving the other intact; the ULG provides an opportunity to know whether (i) the fish has retained the flexibility of differentiation/maturation of gametes during the course of gametogenesis and (ii) the remaining gonad is able to compensate and if so partially or completely the hormone and/or gametogenesis in response to the negative feedback. With the presence of iteroparous and semelparous fishes, the Unilateral Ovariectomy (ULO) may also provide an opportunity to know whether the ULO induces divisions of oogonia with concomitant generation of stem cells. In iteroparous species such as *Heteropneustes fossilis* and *Oncorhynchus mykiss*, eggs are produced in more than one clutch or breeding season, and at a given period, their ovaries consist of follicles at different stages of oogenesis. But in semelparous species like *O. kisutch*, which reproduce only once in a life time, synchronized oogenesis occurs and most of the available follicles are recruited into a single clutch. Following ULO and with the availability of ingredients for egg production, the compensatory ovary may opt for one or more of the following pathways: 1. Produce larger eggs from the existing pool of vitellogenic follicles, 2. Maximize the fecundity by reducing the number of atretic follicles and 3. Increase the number of dividing oogonia with concomitant production of stem cells.

In all fishes, the ULO is typically accompanied by a compensatory course. Firstly, it is accompanied by elevated FSH levels (e.g., Tyler et al., 1997). In iteroparous fishes, the compensatory ovarian hypertrophy is accompanined by the options 2 and 3, i.e., by recruiting additional follicles into the vitellogenesis and reducing the number of atretric oocytes and/or increasing the number of dividing oogonia, implying the incidental generation of additional stem cells (e.g., *H. fossilis*, Goswami and Sundararaj, 1968; *O. mykiss*, Tyler et al., 1994). Contrastingly, the compensatory ovarian hypertrophy of the semelparous *O. kisutch* has limited options of 1 and 2, i.e., producing larger oocytes from the existing pool of follicles and maximizing the fecundity by reducing the number of atretic oocytes (Luckenback et al., 2008). Understandably, the semelparous fish, which are to die after spawning, do not generate additional dividing oocytes and more stem cells.

On the other hand, Unilateral Castration (ULC) of the Atlantic salmon *Salmo salar* and stickleback *Gasterosteus aculeatus* suppresses androgen levels, for instance, the plasma level of 11-KT to 56–63% (Hellqvist et al., 2002). The rapid and complete compensation of androgen levels, especially that of testosterone (see Table 1) observed in the ULC mammals is not found in the breeding males of salmon and stickleback. This observation clearly suggests that the ULC is not regulated by negative feedback.

With availability of more information, a review and critical analysis of regeneration of the ovary and testis in gonadectomized fishes has become necessary. Elevation of the aromatase level in the brain and restoration of masculine behavior of the castrated male show the rapid response in *S. salar* and *Hemihaplochromis multicolor*. However, Mayer et al. (1991) and Reinboth (1972) have remained silent on the regeneration of testis in these castrated fishes (Table 29). Subjected to bilateral gonadectomy, the responses of gonadal regeneration considerably differ one from the other. The castrated *Oncorhynchus nerka* and ovariectomized *Oreochromis niloticus* have regenerated testis and ovary, respectively confirming the decisive sex determination and retention of unisexual potency alone in adults. The grass carp *Ctenopharyngodon idella* also belongs to this group, in which sex is stably fixed and adults retain unisexual potency alone. *C. idella*,

Table 29. Gonadectomy and regeneration in fishes (from Pandian, 2011, modified)

Species/Reference	Reported observations
Betta splendens Noble and Kumpf (1937)* Lowe and Larkin (1975)	65% surviving ovariectomized females regenerated functional testes from the cut end of the remaining oviduct or ovarian wall; crossed with normal females, the regenerated males produced F_1 progenies at sex ratio of 0.43 ♀ : 0.57 ♂
Kaiser and Schmidt (1951)*	Crossed with normal female, the regenerated male sired two broods of six and 102 all-female progenies
Noble and Kumpf (1937)*	Treated with E_2, ovariectomized females failed to regenerate ovary
Gambusia affinis Turner (1967)*	Castrated male apparently regenerated testis, as indicated by the presence of small (4%) testis and complete gonopodium
Oryzias latipes Okada and Yamashita (1994)	Treated with E_2, ovariectomized females failed to regenerate ovary
Carassius auratus Kobayashi M et al. (1991)	Treated with 11-KT, ovariectomized females regenerated testes
Oncorhynchus mykiss Kersten et al. (2001)	Castrated males regenerated testes, irrespective of receiving E_2-treatment. Ovariectomized neomales (XX) regenerated testes
O. nerka Robertson (1961)	Castrated males regenerated functional testes at 'wrong' sites, i.e. adjacent to spermatic duct. Ovariectomized females developed ovaries but with reduced fecundity
Oreochromis niloticus Akhtar (1984)	67% of ovariectomized females regenerated ovaries within 119 days
Ctenopharyngodon idella Underwood (1986)	The gonads including surrounding mesentery were removed. Testes and ovaries were completely regenerated in 91% males and 38% females, respectively
Salmo salar Mayer et al. (1991)	Castrated male increased the aromatase activity in brain on treatment with testosterone or 11-Ketoandrostenedione
Hemihaplochromis multicolor Reinboth (1972)	Castrated male behaved completely as male following injection of testosterone

*cited in Atz (1964)

subjected to MT-treatments between 14–63 dph and between 320–500 dph have not reversed the sex (Hunter and Donaldson, 1983). Even when the entire gonads along with the mesentery have been removed, the female and male grass carps regenerate ovary and testis alone, respectively. Clearly, this experimental evidence reveals that the sex is irrevocably determined at the embryonic/juvenile stage (prior to 14 dph) and not only in their gonads and brains but also the entire fish has been sexualized (i.e., in terms of Zarkower, 2006). In the total absence of gonads including mesentery, the gonads can be completely regenerated from other organ system, i.e., the PGCs, OSCs or SSCs and germ cells supporting cells can be derived from an organ system that is entirely outside the purview of the gonad. *O. niloticus* is also amenable to almost 100% masculization or feminization, provided the respective hormone treatment is synchronized within the labile period of upto 30–45 dph (e.g., Gale et al., 1999). Subsequently, sex is decisively determined and is not altered on gonadectomy. This is true of *O. nerka* also. Castrated genetic male (XY) *O. mykiss* regenerates only the testis, irrespective of receiving E_2-treatment or not. However the castrated neomale (XX) too regenerates testis, indicating the stability of sex of the gonad that has earlier been sex reversed by exposure to exogenous androgen, i.e., it does not re-revert to the genotypic sex during the regenerative process. This implies that the sex of *O. mykiss,* once determined genetically or hormonally, remains stable and adults retain unisexual potency alone. The castrated *Gambusia affinis* regenerates gonopodium even in presence of the small testis, indicating that in *G. affinis* too sex determination is a stable process. Dietary administration of one or other androgen even at relatively high doses to gravid *Poecilia reticulata* and *P. sphenops* have masculinized gestated progenies with undifferentiated gonad but not the gravid females, in which sex is apparently stably fixed (Kavumpurath and Pandian, 1992, 1993; George and Pandian, 1995, 1998). Hence *O. nerka, O. mykiss, O. niloticus, G. affinis P. reticulata, P. sphenops* and *C. idella* belong to a group, i.e., primary gonochores, in which sex is stably fixed by the juvenile stage and their adults retain unisexual potency only, i.e., the germ cells supporting somatic cells of the primary gonochores have retained only unisexual potency by the time they attain the juvenile stage.

On the other hand, the secondary gonochore (see Table 3), *Carassius auratus,* known for its amenability to hormonally reverse gender in either direction, regenerates the testis and ovotestis, following ovariectomy. Clearly, the germ cells supporting somatic cells of secondary gonochores seem to have retained bisexual potency, like the hermaphrodites, which are known to have been derived from secondary gonochores.

Yet, it is difficult to reconcile with the report on the failure of ovariectomized female *O. latipes* and *B. splendens* to regenerate the ovary, even with E_2 administration. They are known for their amenability to reverse gender to either sex on appropriate hormonal treatment during the labile period (e.g., Kirankumar and Pandian, 2003; Balasubramani and Pandian, 2008b). It is not clear whether the adults of these fishes have retained PGCs/OSCs. If the fighterfish has not retained these, how has it regenerated the testis and the male sires male and female progenies, as has been reported by Lowe and Larkin 1975, or only F_1 female progenies, as reported by Kaiser and Schmidt 1951. In the context of PGCs/OSCs/SSCs transplantation research to generate allogenics and xenogenics and their importance in conservation of fishes and aquaculture, it is important to undertake more research to solve the riddles of the observations reported from experimental gonadectomy studies in *B. splendens*.

At this juncture a review of Le Page et al. (2010) proposal of non-sexualization of the brain and the consequent lability of sex differentiation of teleost fishes is to be reconsidered. In agreement with Le Page et al. (2010), almost all the sequential and serial hermaphroditic fishes undergo natural sex change once in one direction and more times in either direction, respectively. Among the simultaneous hermaphrodites too, the rivulus is hatched as a female and becomes hermaphrodite, as it grows (Sakakura and Noakes, 2000); a few gamete exchanging hermaphrodites also become males. Among the sequentials, reports on the sex change by primary males and females are not uncommon . Clearly, the window of brain sexualization covers almost the entire life span and the brains are not sexualized in all the hermaphrodites. Similarly the secondary gonochoric fishes such as the goldfish have also retained bisexual potency during the adult stage, for some of them reverse sex during the adult stage, when subjected to hormonal induction (e.g., goldfish) or social pressure (e.g., paradise fish, see Pandian, 2011). Being a secondary gonochore (see Table 3), the germ cells supporting cells of *C. auratus* have retained bisexual potency. Expectedly, the ovariectomized goldfish, following 11-KT treatment, regenerates the testis, as their germ cells supporting somatic cells have retained bisexual potency (however see Yamaha et al., 2003; see also Table 48, Pandian, 2011). Apparently, the germ cells supporting somatic cells of secondary gonochores, from some hermaphrodites which have evolved, have also retained bisexual potency. To date, the secondary gonochores goldfish, zebrafish, and possibly secondary gonochoric platyfish, paradisefish and checkerboard cichlid are the only gonochoric fishes, which have retained reversible brain sexualization and a long window of brain sexualization (Fig. 56).

On the other hand, Le Page et al. (2010) conclusion that the brains of gonochoric fishes also remain non-sexualized and the window of their brain sexualization includes the matured adult stage, needs further consideration. The available evidences thus far reported do indicate that in these primary gonochores, the window of unsexualized brain lasts only upto the juvenile stage or at the best upto puberty and this duration is synchronized with

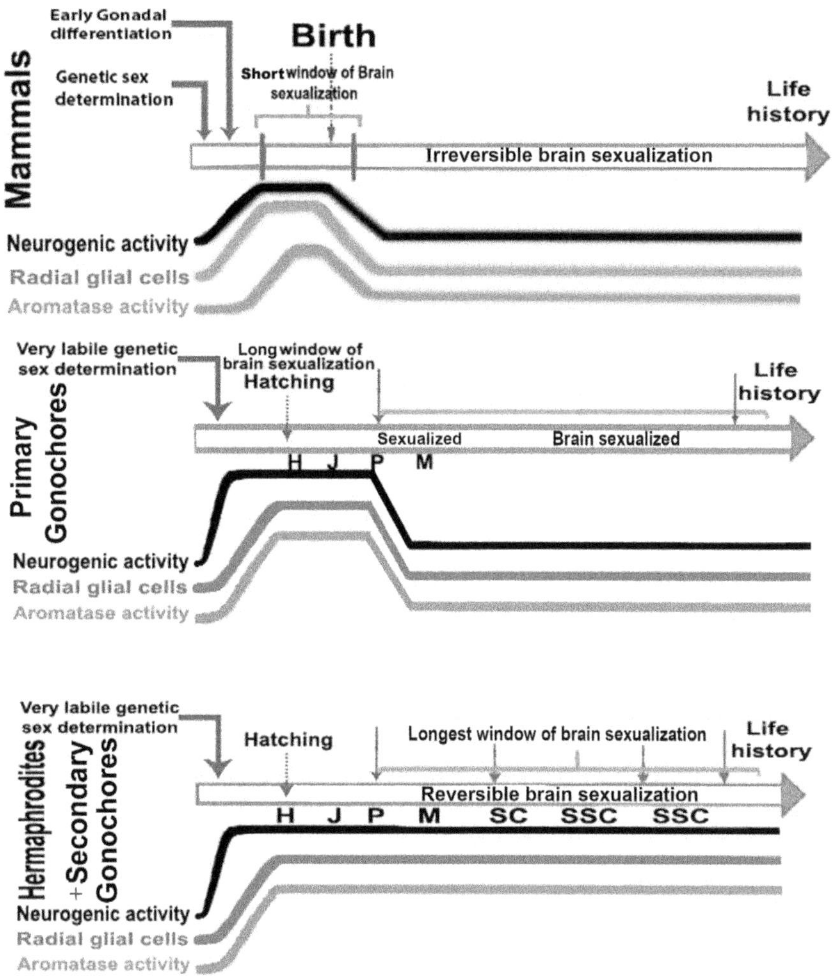

Fig. 56. Proposed ontrogenetic pathways of sex determination and differentiation in mammals (as reported by Le Page et al., 2010, first panel) and the presently proposed pathways for primary gonochoric fishes (second panel) and secondary gonochoric and hermaphroditic fishes (third panel). J = juvenile stage, P = puberty, M = sexual maturity, SC = sex change in sequential hermaphrodites, SSC = sex change in serial hermaphrodites

the so called labile period (Pandian and Sheela, 1995). Firstly all the reported morphotypes undergo ontogenetic change from one morphotype to another but all within one sex namely the male, clearly indicating that the brains of these morphotypes are at least partially sexualized. Secondly, a series of experiments by Yoshizaki and his team has established that bisexual potency of the PGCs, OSCs and SSCs, drawn from mature fishes that have already spawned/milted, express bisexual potency only in the undifferentiated gonad of juvenile recipients, i.e., during the labile period. On transplantation of the SSCs into the sterilized testis containing germ cells supporting cells, the SSCs differentiate into testis only in the Nile tilapia and pejerrey; this indicates that the germ cells supporting somatic cells have lost the bisexual potency and the hormonal climate prevailing in the adult sterilized testis permits the gonad to express unisexual potency of SSCs alone. Administration of one or another androgen even at relatively higher doses to gravid guppy and molly masculinized the gestated progenies with undifferentiated gonad but not the gravid females. Lastly, surgical experiments undertaken by different authors in all the primary gonochores have shown that following the removal of the testis and ovary, the adults regenerate only the testis and ovary, respectively; these observations clearly indicate that as the organ system perhaps including the brain is sexualized, the adult fishes regenerate only the testis following castration and the ovary following ovariectomy. In fact, the grass carp, which is not amenable to the induction of hormonal sex reversal even during the juvenile stage, regenerates the testis and ovary following the removal of the entire testis and ovary including the adjoining mesentery, respectively. Clearly, the sex is decisively determined and the brain of grass carp is sexualized prior to or immediately after hatching. However, many gonochoric teleosts are amenable to hormonal induction of sex reversal, when the hormone is administered during the labile period mostly restricted to the juvenile stage. In them sex is stably determined, even if phenotypic sex is different from genotypic sex, as in the rainbow trout, and their brains are perhaps sexualized at the end of the labile period. It is very likely that the brains of these gonochores are sexualized at the juvenile stage and their window for reversible sex change is certainly closed before the fishes reach sexual maturity (Fig. 56).

4.2 Abnormals and steriles

Among gonochoric species, the presence of abnormal hermaphrodites has been frequently reported. These non-functional or very rarely functional abnormals possess (i) ovotestis (e.g., *Oncorhynchus keta*), or (ii) ovotestis + ovary (e.g., *Salmo trutta*, or (iii) ovotestis + testis (e.g., *O. kisutch*) or (IV) ovary and testis (e.g., *Takifugu rubripes* (Fig. 57). Of these, the incidence of ovotestis ($\approx$ 36%) and ovary + testis (> 33%) is more common (see Atz, 1964). In 1964 Atz summarized more than 90 historical examples. Later Devlin and Nagahama (2002) also summarized the subsequent reports. Pandian (2010) noted that the abnormals are scattered across 23 taxonomic families. For obvious reasons, the incidence frequency is the highest among the commercially important cyprinid species. The reports on incidence of abnormals are mostly limited to the morphological description. Histological

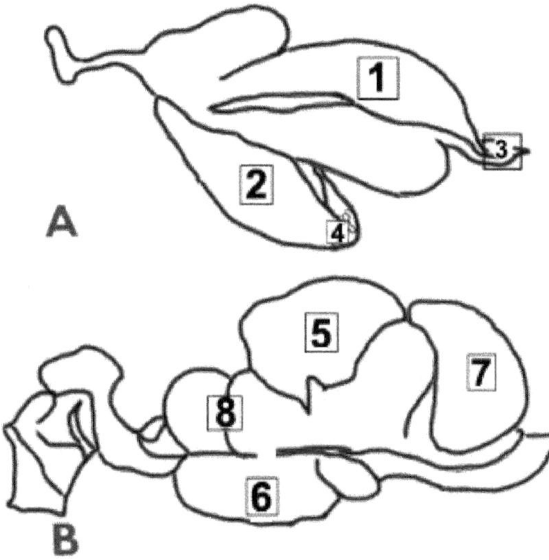

Fig. 57. Diagramatic representation of eight parts of the gonads of hermaphroditic tiger puffers, *Takifugu rubripes* caught from Bungo Channel (A) and in the Japan Sea off Yamaguchi Prefecture (B). 1, 2, 5 and 6 = ovary, 3, 4, 7 and 8 = testis (from Suzuki, 1997)

investigations are limited to *Odontethes bonariensis* (Strussmann et al., 1996) and *T. rubripes* (Suzuki 1997). At the experimental level, Kossman (1971) stripped eggs and sperm form the hermaphroditic carp and successfully fertilized them. While the abnormals seem not to have attracted the attention of physiologists, especially endocrinologists, rare occurrence and non-functional status may have inhibited experimental zoologists from undertaking detailed studies. Nevertheless, the abnormals may increasingly become an important resource for studies on sex differentiation. For Schulz (1996) has reported the regular incidence of 0.02% of abnormals in wild population of the roach *Rutilus rutilus.* Unrestricted and indiscriminate use of pesticides like endosulfan is common in developing countries; the discrete immersion of post-hatchlings of *D. rerio* in 1400 ng endosulfan/l for 14 hours resulted in the production of 44% abnormal hermaphrodites (Balasubramani and Pandian, 2008a). Secondly, the genome of non functional/inviable abnormals can be recovered using the newly developed technique of Shimada and Takeda (2008). These abnormals may provide excellent opportunities for a better understanding of the high plasticity encountered in sexuality of fishes.

Steriles: Contrary to the abnormal hermaphrodites, steriles are those with gonads of reduced size, where gametogenesis has been blocked (Pifferer et al., 1994). Whether the steroidogenesis is blocked or not is unknown as yet. C Strussmann introduced sperm cells into sterile *Odontesthes argintinensis* and found them to be alive until the 10th week after introduction (see Pandian, 2003). Apparently, Leydig cells in this sterile fish may have been active providing nutrients to the transplanted sperm. Attempts have been made to induce sterility in commercially important fishes under the impression that these 'broilers' may grow faster. As indicated elsewhere, sterile gonads have not yet attracted electron-microscopic studies (cf Fig. 41,54) to provide details on the presence of PGCs, OSCs SSCs and germ cells supporting somatic cells. Dorsch et al. (2004) have identified 20 male sterile mutants in zebrafish *shooting banks* (shb^{p12idc}) and are known to cause a defect in sperm motility. Transmission and scanning electronmicroscopic investigations of the spermatozoa of *Cyprinus carpio* have shown 9.5, 5.3 and 2.4% of biflagellar, triflagellar and double-headed abnormal spermatozoa. These sperms also show differences in ultrastructure, motility and relative DNA content, compared to normal sperms (Psenicka et al. 2009). Reports on the natural occurrence of steriles are indeed very rare. But they are artificially produced following hybridization, and ploidy and hormonal inductions. For details on ploidy and hormonal induction of steriles, Pandian and Koteeswaran (1998), Pandian and Sheela (1995) may be consulted. Artificial hybridizations have been induced in more than 300 species (see Argue and Dunham, 1999; Bartley et al., 2001; Pandian, 2011); however, viable and fertile progenies are generated only in very few instances. It is now established that hybridization results in male sterility but triploidization in female sterility. Triploid males are hormonally fertile but cytologically sterile, while female triploids are both hormonally and cytologically sterile. Corresponding information is needed for hybrid male sterility.

4.3 Mutants

Mutants provide a crucial tool to investigate the role of genes in a biological process. Collection of mutants in a process can be the key to elucidate molecular mechanisms or pathways controlling the process. The growing number of sequenced genomes and overall genome resources available for fishes have made them an important tool for investigations of human genetics, development and diseases. For instance, experimental studies of spinal deformities in fishes remain an unexplored resource, capable of answering questions of vertebrate development, basic stability and deformity (Gorman and Breden, 2007). Among the vertebrate species, in which the mechanisms of embryogenesis are currently investigated, the zebrafish offers several advantages. Generation time of three months, and regular weekly spawning of about 100 eggs undergoing synchronous development outside the mother are some advantages (Solnica-Krezel et al., 1994). Besides the almost transparent embryos allow monitoring the early development of the brain, eye, trunk musculature, gut, heart and nervous system without having to dissect them. Analysis of their mutations has demonstrated the value of the system for genetic dissection of vertebrate development and diseases (Talbot and Hopkins, 2000).

In vertebrates, maternal genes direct fertilization, egg activation, the first cell division(s) and initiation of zygote transcription (Dorsch et al., 2004). In amphibians and fishes, the Animal-Vegetal pole (AV) or prospective anterior-posterior axis is already established during oogenesis, as evidenced by the asymmetric localization of germinal vesicle (King et al., 1999; Suzuki et al., 2000). The dorso-ventral axis, established during embryonic cleavage, also relies on maternal gene products. Thus the maternal genes and their products lay the foundation for embryonic development, on which zygotic factors build upon later. For example, Dorsch et al. (2004) have identified 68 maternal effect mutants in zebrafish. Of them 15 are described to control the formation of blastodisc, embryonic polarity, initiation of cell cleavage and division (Table 30). Incidentally, Araki et al. (2001) have shown that

Table 30. Selected maternal genes and their role in early embryonic development in zebrafish (compiled from Dorsch et al., 2004 and others)

Classes of mutants and their location in chromosomes (chr)	Reported observations
	Opaque egg
Ruehrei ^{p25ca} on chr 6 *over easy* ^{p37ad} *sunny side up* ^{p144de} on chr 22 *souffle* ^{p96rc} on chr 16	These eggs fail to segregate cytoplasm to animal pole and to form bastodisc. The failure is due to defective oogenesis during stage IV
	Egg activation and cytoplasmic segregation
Jumpostart ^{p108re} *p11cv* *Emulsion* ^{p41p1} on chr 4 dp *14nb*	Egg activation includes fusion of cortical granules to the egg membrane. Release of their contents results in chorion hardening One or more of these events fail. Involved in activation and formation of microfilament required for blastodisc formation
	Animal–Vegetal polarity
Buck ball ^{p106re} *p6cv*	Wrong localizations of *cyclinB* mRNA in animal pole and *Bruno–like* in vegetal pole. Responsible for streaming of cytoplasm and consequent cytoplasmic segregation to the animal and vegetal poles. Cold shock affects cytoplasmic tubulin formation
	Failure to initiate cleavage
atomos ^{p71fm} on chr 9 *indivisible* ^{p15dia} *irreducible* ^{p15mf} on ch 20	Responsible for initiation of cytokinesis and karyokinesis. *Indivisible eggs* are fertilizable but do not initiate cleavage
	Incomplete cellurization
Cellular island ^{p63cd} *cellular atol* ^{p57mfa} on chr 22	Responsible for completion of cleavage. Mutants cause failure of cytokinesis but karyokinesis proceeds leading to the formation syncytium, as in *cei* nuclei

screening of haploid medaka is useful to identify drastic morphogenic mutants showing dorso-ventral polarity and antero-posterior pattern of the Central Nervous System (CNS) but it is not suitable to identify genes controlling regional polarity in organs and boundary formation between tissues.

Large scale mutational screens of the medaka have identified a large number of genes that are zygotically required for normal embryonic development (Furutani-Seki et al., 2004; Morinaga et al., 2004). These zygotic mutants are now recognized as powerful tools for the study of the genetic cascades involved in vertebrate development (Shimada and Takeda, 2008). In most animals, the genomes of fertilized embryos are transcriptionally silent until a specific stage of early development; the maternally supplied products, synthesized and accumulated during oogenesis, may attenuate the phenotype of a zygote mutant fully or partially, repairing the lost gene(s). To understand the early functions of such mutated genes, a mutant, that lacks both maternal and zygotic gene products, referred to as a maternal-zygotic mutant, is required.

Taking advantage of the sexually differentiated but sterile hybrid between *Oryzias latipes* and *O. curvinotus*, Shimada and Takeda (2008) developed a method of using naturally available sterile ovary and testis of these interspecific hybrids as recipients, in which germ-line is restored by transplanting PGCs of heterozygous MZ mutant for *fgfr1* (lethal heal/fibroblast growth factor) (Fig. 58). The resulting mutant reveals the precise role of *fgfr1*-mediated signaling in the early embryos. Out of 174 transplanted recipient embryos, 30% are found to have donor PGCs that have successfully migrated and colonized the recipients' gonad. Among the 20 recipient females, 60% recovered their fertility and produced donor-derived eggs. Clearly, the female hybrid sterility can be attributed to a defect in the germ cell lineage and the somatic cells in the hybrid ovary have retained the ability to support oocyte development. However, fertility is recovered in the female transgenic medaka. A similar research has been performed in the danios, in which fertility is recovered in males but not in females (Wong et al., 2010).

There are also examples for artificial induction of mutants resulting in alterations in the dorso-ventral axis. Thermal shock treatment affects and prohibits the release of the second polar body or the first cell division (Pandian, 2011). The treatment may alter the formation of tubulin in egg cytoplasm, which plays an essential role in segregation of chromosome sets to daughter cells. The microtubules also regulate the early cellular differentiation of blastomere from vegetal to the animal hemisphere of embryos or *vice versa*. From eggs that underwent heat shock for 30 to 90 seconds at 40°C, a high percentage of goldfish embryos were shown to receive reduced signal or no signal of *gooscoid (gsc)* mRNA (Fig. 59),

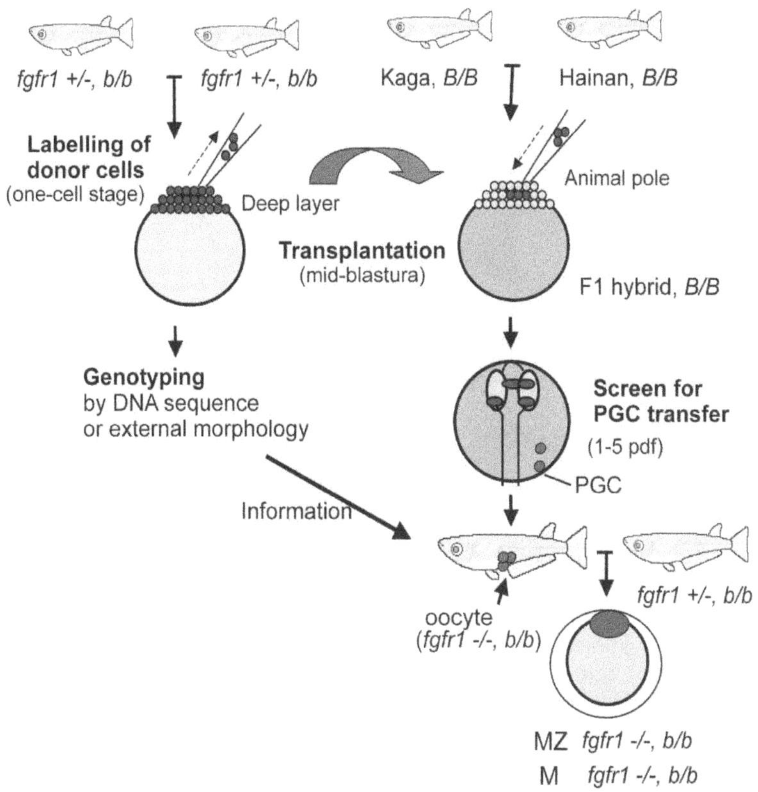

Fig. 58. Schematic representation of the production of a maternal zygotic (MZ *hdf/fgfr1*) mutant by germ-line replacement. At the midblastula stage, the donor cells are transplanted from the deep layer of the blastoderm into the animal-pole region of a hybrid host embryo at the same stage. Successful PGC transfers are scored at segmentation stages from 1 to 4 dpf. The host embryos with *fgfr1* -/- PGCs were mated with *fgfr1* +/- males to obtain MZ *fgfr1* -/- embryos (from Shimada and Takeda, 2008, modified)

resulting in both polar body retension and suppression of dorso-ventral differentiation (Yamaha et al., 2002).

Influence of maternal genes seems to last even beyond the early embryonic development. For example, from a construction of a maternal EST library containing 2,431 sequences, Mommens et al. (2010) have identified two genes *HCOO353* and *HHCO1517* that are positively correlated with the hatching process and the third one *HHCOO255,* that is negatively correlated with the percentage of normal blastomeres. Apparently, these maternal genes determine the quality of eggs. Incidentally, the non-symmetrically cleaving eggs used as a predictor of egg quality in *Limanda ferruginea* (Avery and Brown, 2005) and *Gadus morhua* (Hansen and Puvanendran, 2010) may represent mutants of maternal genes that determine egg quality.

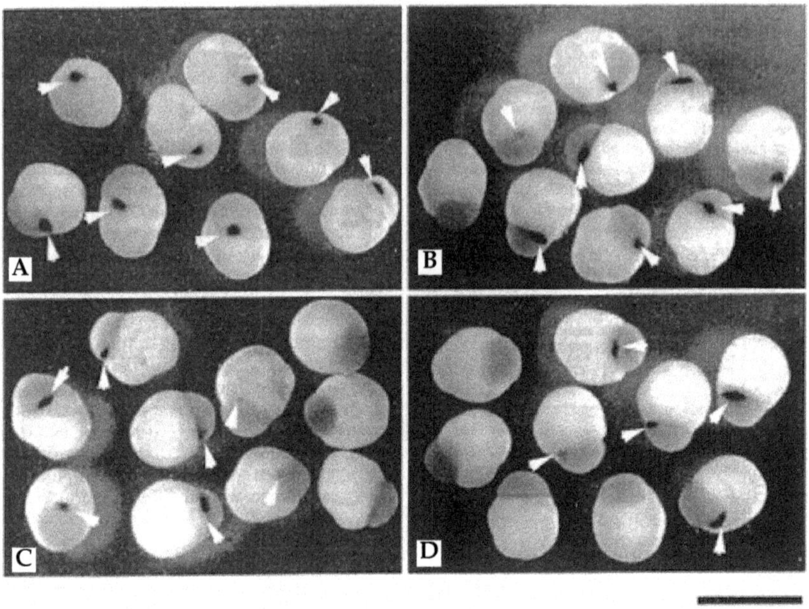

Fig. 59. Expression of *goosecoid* gene at 12 hour post-fertilization in (A) control embryos of goldfish and (b-d) heat shock—treated embryos. Embryos were treated at 40°C for (B) 30 seconds, (C) 60 seconds and (D) 90 seconds at 5 minutes post-fertilization. Arrow heads indicate the reception of signal of *gsc*. Bar = 1 mm (from Yamaha et al., 2002)

In the context of global warming, Ytteborg et al. (2010) exposed during the period from the egg stage to 60 g body size of *S. salar* to low temperature at 10°C and high temperature at 16°C. In the fish exposed to higher temperature, faster growth disrupted the differentiation of bone and cartilage cells resulting in a significantly higher frequency of deformities in spinal column than in those reared at 10°C. The increased risk of developing vertebral deformities was linked to an altered gene transcription, especially the down-regulation of extracellular matrix (*ECM*) genes like *Colla1, Osteocalcin* and *decorin*. Exposing the 5- and 10- d old fry of *O. mossambicus* to selected temperatures (20–33°C) for five days, Wang and Tsai (2000) also noted that exposure of 5-day old fry at high temperature (28–32°C) alone caused a high percentage of deformity. Gavaia et al. (2002) recognized high number of malformations both in the caudal complex and vertebral column of *Solea senengalis*. About 44% of the individuals showed at least one deformity. Of them 28% deformities were in vertebral column.

The study of Gorman et al. (2007) on the mutant guppy *P. reticulata Curveback* syndrome stands as a unique example for a better understanding of the spinal stability and deformity. Due to extreme phenotypic variability, very little progress has been made to identify the etiology of idiopathic scoliosis in humans. Observed parallels between the guppy *Curveback* syndrome and human idiopathic scoliosis suggest that the guppy model is an unexploited resource for the identification of primary etiological factors involved in curvature. Computed tomography of the guppy has revealed that vertebral breakage or fusion is not associated with the *Curveback* syndrome. In the guppy, the curves are classified into four types: 1. Nearly curved, 2. Slightly curved, 3. Moderately curved and 4. Progressed curve. The curves of greater amplitude begin within 4 dpb (days post birth), while curves that are initiated after 7 dpb tend not to progress fast in type 2. Progression of curves is periodic, oscillating between periods of increase or decrease and relative stability. Selected crossing experiments of the guppies with different types of curves described above have revealed that the gene controlling *Curveback* is autosomal recessive. The proportion of affected *Curveback* individuals in the lineage is increased in successive generations with inbreeding, for example, from 50% of F_2 offspring curvature during development to 58% in F_3 generation. However, there is a female bias for curves of high magnitude, i.e., 15% of females have severe (type 5) curvature, compared to 4% of males. Incidentally, at least one skeletal trait, the size of the opercle bone is mapped to the sex chromosome of three spine stickleback (Kimmel et al., 2005).

Led by Morinaga, a team of 18 Japanese scientists undertook a large scale screening of medaka mutants; in them, the abnormal presence and/or distribution of germ cells were detected by in *situ* hybridization of *olvas*, the vasa of medaka (Morinaga et al., 2004). They described 16 mutants caused by 13 genes, which are classified under four groups. Group 1 consists of four mutants; three of them are recessive genes that increase the number of germ cells. Of them *seitaka* (*sei*), *kongara* (*kon*) are not viable but *totora* (*tot*), inducing gonadal hypertrophy, is viable. In Group 2, *zanzai* (*zei*) mutant decreases the number of germ cells. Group 3 includes mostly semi-dominant or recessive mutants; they are *ekou* (*eko*), *eki* (*eki*), *shitoku* (*sht*) and *anokuba* (*ano*) and are associated with irregular clustering of germ cells, resulting in testicular hypertrophy, gonadal dysgenesis, tumorigenesis, short body and twisted trunk, respectively. Group 4 consists of *arara* (*arr*), *hyou* (*hyo*), *mizore* (*mzr*), *hadare* (*hdr*) and *fubuki* (*fbk*) and are recessive or semi-dominant genes responsible for fragmented clustering of germ cells. Of them, *hyo* alone is viable. Among all of them, the most interesting one is the *tot* mutant, in which the hypertrophic gonad is filled with immature oocytes. This shows that oogenesis is arrested prior to the vitellogenic stage, while the germ cell proliferation continues. The genetic male *tot* mutants possess *Dmy* sequence, indicating that the *tot* phenotype develops independent of the genetic sex.

The nitroreductase (*Ntr*) gene of *Escherchia coli* encodes an enzyme that converts prodrugs like metronidiazole (Met) to cytotokins. Hsu et al. (2010) have generated transgenic zebrafish, in which the *eGFP: Ntr* is expressed under the control of the spermatocyte specific Asp, old and sam promoters. The transgenic females are fertile; the males have small testes that are virtually devoid of germ cells and suffer sterility but otherwise appear normal. Hence the zebrafish transgenic line shows a perfect testis specificity of inducible male sterility. Hsu et al. (2010) claim that their 'infertile breeding technology' has practical application in controlling undesired genetically modified fish, and meets the standards of biological and environmental safety.

Whereas the earlier described screens of fishes have concentrated on early developmental mutations, the study of Bauer and Goetz (2001) demonstrates that the mutagenesis can be used to understand the process of gonadal differentiation and gametogenesis in adults. A mutagenesis screen has been made on adult zebrafish subjected to N-ethyl-nitrosourea (ENU) as a mutagen and a F_2 crossing scheme has also been used to generate F_3 homozygous mutants. From a screening of 125 mutagenized genomes generated in 81 families, 11 mutants, that produce visible phenotypes in only one sex per family, have been identified. The males suffer three mutations alone but the females as many as eight. These mutations alter spermatogenesis with the formation of spermatogonia or spermatocytes alone and different aspects of ovarian histology. The most interesting observation made by Bauer and Goetz (2001) is that in these males, the body size may not affect testis development (cf Loukovitis et al., 2011) but the ovarian development may directly be linked to body size.

The observations of Feistma et al. (2007) show that thanks to mismatch repair genes, multiple solutions are present for problems of meiotic crossover/ segregation difficulties. Male *Mlh1* mutant zebrafish are sterile and display an arrest in spermatogenesis at metaphase 1 (Fig. 60), resulting in increased testis weight due the accumulation of prophase 1 spermatocytes and other cytological problems including increased apoptosis. Contrastingly, females are fully fertile but their progeny shows high incidence of malformations (Fig. 61) and suffers mortality within the first few days of development. SNP based chromosome analysis revealed that missegregration of chromosomes at meiosis 1 causes aneuploidy. However a small percentage of surviving progenies are triploids consisting of both two sets of maternal chromosomes and one set of paternal chromosomes. The adult triploid males have low fertility, due to synapsis problems or remain sterile. The frequency triploid progeny of *Mlh1* mutant females is much higher than that expected from random chromosome segregation. Hence there is a way, though narrow one, for the *Mlh1* mutants to escape into triploids and survive.

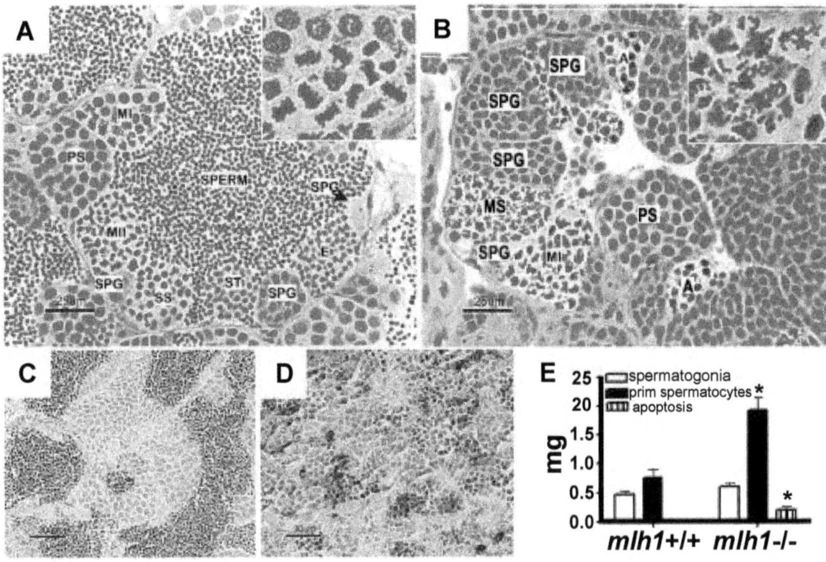

Fig. 60. Cross sections of seminiferous tubule in (A) *mlh1*$^{+/+}$, (B) *mlh1*$^{-/-}$ zebrafish, showing spermatogenic cysts with different types of germ cells, spermatogonia (SPG), primary spermatocytes (PS), first meiotic division (MI), secondary spermatocytes (SS), second meiotic division (MII), spermatids (ST), spermatozoa (SPERM), and apoptotic spermatocytes (A). Note the absence of postmeiosis I stages and the presence of apoptotic spermatocytes with strongly condensed nuclei in *mlh1*$^{-/-}$ testis. Inset shows a magnified view of the first meiotic division. In the wild type, chromosomes are aligned at the cell equator just prior to division. In the mutant, chromosomes are randomly distributed throughout the nucleus. TUNEL staining of (C) *mlh1*$^{+/+}$ and (D)*mlh1*$^{-/-}$. High numbers of apoptotic spermatocytes are seen in mutant testis. Wild-type testis shows a very low incidence of apoptosis. (E) Morphometric analysis of testes sections showing increase in amounts of spermatocytes and apoptoic cells, but not of spermatogonia. (from Feitsma et al., 2007, with permission by the Genetic Society of America)

Color image of this figure appears in the color plate section at the end of the book.

Kishi et al. (2008) screened mutagenized zebrafish embryos for the altered expression of a stress biomarker, senescence-associated β-galactizidase (*SA*-β-*gal*). For example, the quantitative analysis of *SA*-β-*gal*, a biomarker of ageing, shows a near linear increase in *SA*-β-*gal* activity with the age of zebrafish (Fig. 62). From a pool of 306 retirovirus-insertional mutants zebrafish, Kishi et al. selected two candidate genes. Of these, one is known to regulate life span, the *spinster*, and the other *telomeric repeat binding factor 2* (*terf2*) gene, which plays a role in telomere protection and telomere length regulation. Both *spinster* and *terf2* mutants are embryonic lethal but their heterozygous progenies are viable and show an accelerated appearance of ageing symptoms including accumulation of lipofusion, a

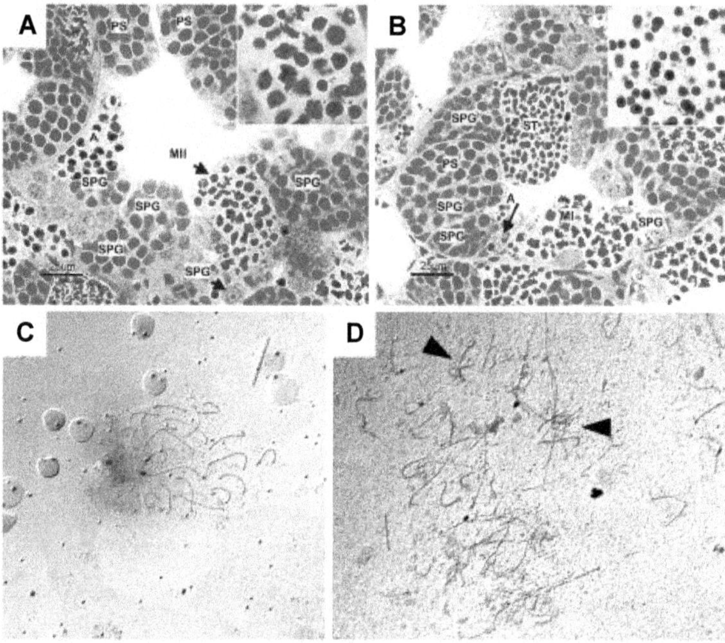

Fig. 61. Cross sections of seminiferous tubules in (A) infertile and fertile (B) triploid zebrafish. For expansion of abbreviations, see Fig. 4.3. Inset in (A): magnification of second meiotic division; note the concomitant presence of secondary spermatocytes and spermatids (F) and G). Synapsis in triploid spermatocytes (D) is strongly disorganized as compared to that in wild type (C). Irregular synapsis is indicated by arrow heads (from Feitsma et al., 2007, with permission by the Genetic Society of America)

Color image of this figure appears in the color plate section at the end of the book.

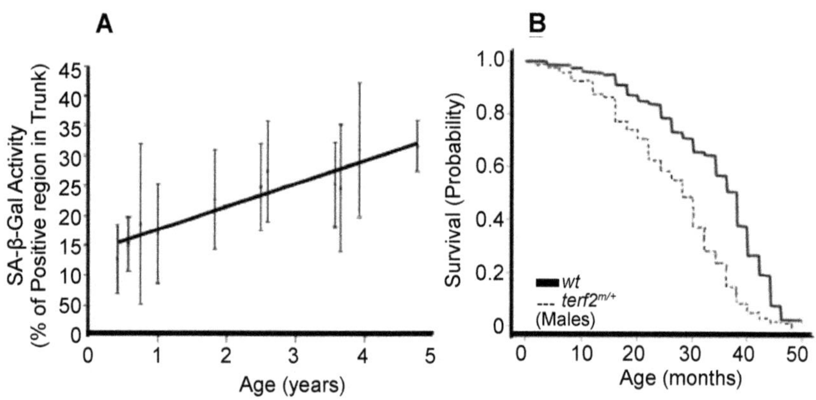

Fig. 62. A *SA-β-gal* activity in the trunk skin of the adult zebrafish increases with age. B Note the shorter life span of heterozygous *terf2*$^{m/+}$, in comparison to *wt* zebrafish (from Kishi et al., 2008)

biomarker of a short life span. Mutant *terf2* zebrafish has a smaller eye and high *SA-β-gal* activity in the brain and spinal cord. Not surprisingly, *terf2m/+* mutants, for example male, have a 5–10 months shorter life span.

5

Highlights and Directions

Le Page et al. (2010) have indicated that in mammals sex is decisively determined by a single *Sry* gene located on a morphologically distinguishable sex chromosome. In fishes sex determination remains indecisive, as it is determined by many genes (e.g., Hale et al., 2010; Loukovitis et al., 2010; Shirak et al., 2006) located on different and not readily distinguishable chromosomes (e.g., Bradley et al., 2011); some of these genes are drastically influenced by endocrines and similar chemicals (Pandian et al., 2012) as well as biotic (social) and abiotic (e.g., temperature, Pandian et al., 2012) factors. In this context, it is interesting to note that many leading scientists (e.g., Herpin and Schartl, 2011) have concluded that many (sex determining) genes at the top of the genetic cascade are not conserved, while the downstream genes are homologous in a much broader spectrum of fishes. Hence the need for genome-wide linkage analysis and search for more than one sex determining gene residing on different chromosomes of fishes is emphasized.

Le Page et al. (2010) have proposed that non-sexualization of the brain is the root cause for the plasticity in sex differentiation in fishes. However, this may hold true only for all sex changing hermaphrodites and secondary gonochores alone. The brains of primary gonochoric fishes are indeed sexualized on or before the juvenile/pubertal stage. This conclusion is based on: (i) all the phenotypic morphotypes undergo intrasexual ontogenetic change suggesting brain sexualization in them, (ii) PGCs from embryos and OSCs/SSCs from differentiated gonad drawn from mature adults that have already spawned/milted, when transplanted into undifferentiated gonads of recipients, express bisexual potency. However, the transplantation of SSCs into sterilized adult testis containing germ cells supporting cells differentiate into testis alone; this clearly indicates that the germ cells supporting somatic cells have lost the bisexual potency, i.e., brain sexualization has decisively acted on germ supporting cells but not on SSCs/OSCs, (iii) From their experiments on hormonal sex reversal, Nakamura et al. (2003) have reported that both germ cells and germ cells supporting cells in the differentiated ovaries have retained bisexual potency. Hence there is a need for further research to confirm whether the germ cells supporting cells of primary gonochores retain bisexual potency. Notably, the experiments undertaken by Nakamura et al. (2003) have been limited to the juvenile stage (i.e., labile period) during which the brain is not yet sexualized. Incidentally, the dietary administration of androgens to gravid females induced sex reversal only in the progenies but not in the adult female guppy and molly with differentiated ovary (Kavumpurath and Pandian, 1992, 1993; George and Pandian, 1995, 1998), (iv) Gonadectomized fishes regenerated testis only, when castrated and ovary only, when ovariectomized, indicating that not only brains of these primary gonochores but also the entire organ system in these fishes have been sexualized. Interestingly, a clear sexually dimorphic expression in muscles of adult male and female guppy *Poecilia reticulata* has been reported (Shen et al., 2007). At least one skeletal trait, the size of the operclar bone has also been mapped to the sex chromosome of three spined stickleback *Gasterosteus aculeatus* (Kimmel et al., 2005). Hence more and more evidences from different spheres are forthcoming to confirm that the brain and organ systems of adult primary gonochores are sexualized. An exception seems to be *Betta splendens,* in which contradictory results have been reported indicating that more gonadectomy studies are required, especially in anabandits like *B. splendens, Trichogaster* and *Macropodus.*

When recognizing the two different ontogenetic pathways of sex differentiation and naming them as primary and secondary gonochores, it is not clear whether Yamamoto (1969) was aware of the possibility that hermaphrodites originated from secondary gonochores. For this classification is a fundamental contribution of Yamamoto; it brings the primary gonochores on one side, and the secondary gonochores and hermaphrodites on the other. The brains of the former are sexualized and their window for sex change is closed prior to puberty but the reverse is true for the latter. The gonads of almost all the types of secondary gonochores commence as differentiated ovary or ovotestis, which are similar to those in hermaphrodites (see Figs. 3,4). From these differentiated ovaries/ ovotestes of the secondary gonochores, 50% of them are subsequently de-differentiated and re-differentiated into testes during the post-hatching stage in *Danio rerio* or just prior to puberty in *Dascyllus flavicaudus.* Thus the mechanism for re-differentiation, a stepping stone required for natural sex change in hermaphrodites has been established in these secondary gonochores. This stepping stone has been pushed further to mid-adult stage in the hermaphrodites. This is how hermaphrodites may have arisen from secondary gonochores. It must be noted that after all the processes de-differentiation and re-differentiation in the same original genetic sex of the gonads do occur in seasonal/annual spawners repeatedly. But the most crucial step is the genes that are responsible for switching the redifferentiation process quite to the opposite sex. Hence it may prove to be exciting in undertaking comparative studies on some of these secondary gonochores, hermaphrodites and hermaphrodites that have returned to gonochorism (e.g., *Leptoscarus vaigiensis*) to identify the genes that are responsible for switching the re-differentiation process on and off.

Incidentally, diandric hermaphrodites have undelimited type gonad. The testis is reported to be derived from its wall. Most available reports are based on histological studies of the ovarian wall of the Japanese diandric fishes. From the picture and description of these ovarian walls, the existence of germ cells supporting cells like the Sertoli cells and Leydig cells is not discernible. An electron microscopic study like the one made for the sterile gonad of tilapia, is desirable.

The process of sex differentiation has been described considering gonochores and hermaphrodites separately. For gonochores, with the exceptions of a few (e.g., *Macropodus opercularis, Crenicara punctulata,* see Pandian, 2011) do not undergo natural sex change as adults. However, hermaphrodites do undergo natural sex change once in a single direction or more than once in either direction; the exceptions may be a fractions of females in *Pagrus pagrus* and males in *Diplodus vulgaris* (see Pandian, 2010). Nevertheless, our description of sex differentiation genes and sex steroid receptor genes has clearly shown that they are not only common to both of them but also homologous, confirming that the hermaphrodites have originated from secondary gonochores and have retained the same genetic mechanism of gonadal differentiation.

Though much is known about the phenotypic morphotypes (e.g., Pandian, 2010) and their endocrine profiles (e.g., Oliveira, 2006), this book recognizes that all phenotypic morphotypes reported among gonochores are *intrasexual,* i.e., fall within a single sex. But those among hermaphrodites are *intersexual,* i.e., they cross the borders from one sex to the opposite; for example, primary male to female and vice versa. More reports are required to confirm this conclusion.

Genotypic morphotypes include the jack and hooknose among salmonids, parental and cuckolder among centrarchids and red and yellow morphs among cichlids. The XY hooknose is known to sire both hooknose and jack offspring. It is not known whether jacks too sire jacks and hooknoses. Also it is not known whether the same is true of the centrarchid and cichlid genetic morphotypes, as has been observed for the hooknose of salmonids.

To assess the reproductive capacity of fishes, the conventional method is to estimate the gonado-somatic index. The present trend is to estimate Fecundity (F). Because F is related to the volume of body cavity available to accommodate the ripe ovary, F is estimated mostly in relation to body Length (L) and sometimes to body Weight (W). However, these simple exercises, when designed more meaningfully to cover the entire life span of the fish, may have far-reaching implications to aquaculture and broodstock management as wells reproductive senescence, which, in turn, reduce or cease stem cell production. Some of these estimates made for the first time in this presentation identify the optimum female size for broodstock and show that smaller fish with a short life span, which are likely to be subjected to high predation, allocate more of its resource for reproduction, whereas the larger and long living fishes allocate for somatic growth. Thereby it is possible to identify the right candidate species for aquaculture of 'bait fish' and food fish. More interestingly, these estimates, covering the entire life span may also provide valuable information on reproductive senescence and stem cell production. Conventional fishery biologists may have to estimate F of commercially important fishes covering the entire life span.

Many authors have reported the appearance of mutants during the embryonic and post-hatching stages. The study on *curveback* syndrome of mutant guppy stands as a unique example of the spinal stability and deformity. There is an urgent need to subject mutants for an indepth investigation for a better understanding of genetic sex differentiation in fishes, as models for human diseases and genetic therapy.

References

Abrams, EW, Mullins, MC. 2009. Early zebrafish development: It's in the maternal genes. Curr Opin Genet Dev, 19: 396–403.

Aida, T. 1936. Sex reversal in *Aplocheilus latipes* and a new explanation of sex differentiation. Genetics, 21: 136–153.

Ajith Kumar, TT, Balasubramanian, T. 2009. Broodstock development, spawning and larval rearing of the false clownfish, *Amphiprion ocellaris* in captivity using estuarine water. Curr Sci, 97: 1–3.

Akhtar, N. 1984. Anesthesia, abdominal surgery, efficacy of ovariectomy and subsequent androgen treatments in inducement of sex reversal in *Tiliapia nilotica*. PhD. Thesis, Auburn University, Auburn, USA.

Alam, MA, Nakamura, M. 2007. Efferent duct differentiation during female to male sex change in honey comb grouper *Epinephelus merra*. J Fish Biol, 71: 1192–1202.

Alam, MA, Kobayashi, Y, Hirai, T, et al. 2010. Isolation, characterization and expression analysis of FSH receptor in protogynous grouper. Comp Biochem Physiol Part A, 156: 364–371.

Allender, CJ, Seehausen, O, Knight, ME, et al. 2003. Divergent selection during speciation of Lake Malawi cichlid fishes inferred from parallel radiations in nuptial colouration. Proc Natl Acad Sci USA, 100: 14074–14079.

Allsop, DJ, West, SA. 2003. Constant relative age and size at sex change in sequentially hermaphroditic fish. J Evol Biol, 16: 921–929.

Alvarez, MC, Otis, J, Amores, A, et al. 1991. Short term cell culture technique for obtaining chromosomes in marine and fresh water fish. J Fish Biol, 39: 817–824.

Amanze, D, Iyengar, A. 1990. The micropyle: a sperm guidance system of teleost fertilization. Development, 109: 495–500.

Amberg, JJ, Goforth, R, Stefanavage, T, et al. 2009. Sexually dimorphic gene expression in the gonad of river shovelnose sturgeon (*Scaphirhynchus platorhynchus*). Fish Physiol Biochem, 36: 923–932.

Aoki, Y, Nagao, I, Saito, D, et al. 2008. Temporal and spatial localization of three germ line-specific proteins in medaka. Dev Dyn, 237: 800–807.

Araki, K, Okamoto, H, Graveson, AC, et al. 2001. Analysis of haploid development based on expression patterns of developmental genes in the medaka *Oryzias latipes*. Develop Growth Differ, 43: 591–599.

Argue, BJ, Dunham, RA. 1999. Hybrid fertility, introgression and backcrossing in fish. Rev Fish Sci, 7: 137–195.

Artieri, CG, Mitchell, LA, Ng, SSH, et al. 2006. Determination of the sex determining locus of Atlantic salmon (*Salmo salar*) on chromosome 2. Cytogenet Gen Res, 112: 152–159.

Asoh, K. 2003. Gonadal development and infrequent sex change in a population of the humbug damselfish *Dascyllus aruanus* in continuous coral-cover habitat. Mar Biol, 142: 1207–1218.

Asoh, K. 2004. Gonadal development in the coral-reef damselfish *Dascyllus flavicaudus* from Moorea, French Polynesia. Mar Biol, 146: 167–179.

Asoh, K. 2005. Frequency of functional sex change in two populations of *Dascyllus melanurus* confirms to a prediction from sex allocation theory. Copeia, 2005: 732–744.

Asoh, K, Kasuya, M. 2002. Gonadal development and mode of sexuality in a coral-reef damselfish *Dascyllus trimaculatus*. J Zool (Lond), 256: 301–310.

Asoh, K, Yoshikawa, T. 2003. Gonadal development and an indication of functional protogyny in the Indian damselfish *Dascyllus carneus*. J Zool (Lond), 260: 23–39.
Asoh, K, Yoshikawa, T, Kasuya, M. 2001. Gonadal development and non-functional protogyny in a coral-reef damselfish *Dascyllus albisella* Gill. J Fish Biol, 58: 1601–1616.
Atz, JW. 1964. Intersexuality in fishes. pp 145–232. In: CN Armstrong, AJ Marschall (eds). Intersexuality in Vertebrates including Man. Academic Press, London, UK.
Augenlicht, LH, Kobrin, D. 1982. Cloning and screening of sequences expressed in a mouse colon tumor. Cancer Res, 42 : 1088–1093.
Avery, TS, Brown, JA. 2005. Investigating the relationship among abnormal patterns of cell cleavage, egg mortality and early larval condition in *Limanda ferruginea*. J Fish Biol, 67: 890–896.
Avise, JC, Mank, JE. 2009. Evolutionary perspectives on hermaphroditism in fishes. Sex Dev, 3: 152–158.
Balasubramani, A, Pandian, TJ. 2008a. Endosulfan suppresses growth and reproduction in zebrafish. Curr Sci, 95: 883–890.
Balasubramani, A, Pandian, TJ. 2008b. Effect of norethindrone on masculinization of fighting fish. Curr Sci, 95: 1446–1453.
Baroiller, JF, Guiguen, Y, Fostier, A. 1999. Endocrine and environmental aspects of sex differentiation in fish. Cell Mol Life Sci, 55: 910–931.
Baron, D, Houlgate, R, Fostier, A, et al. 2005a. Large scale temporal gene expression profiling during gonadal differentiation and early gametogenesis in rainbow trout. Biol Reprod, 73: 959–966.
Baron, D, Fostier, A, Breton, B, et al. 2005b. Androgen and estrogen treatments alter steady state messenger RNA (mRNA) levels of testicular steroidogenic enzymes in the rainbow trout *Oncorhynchus mykiss*. Mol Reprod Dev, 71: 471–479.
Barrero-Gil, J, Thomas, P, Piferrer, F. 2004. Estrogen binding sites in the ovary of the European sea bass (*Dicentrarchus labrax*). Fish Physiol Biochem, 28: 369–370.
Bartley, DM, Rana, K, Immink, AJ. 2001. The use of interspecific hybrids in aquaculture and fisheries. Fish Biol Fish, 10: 325–337.
Bass, AH. 1996. Shaping brain sexuality. Am Sci, 84: 352–363.
Batargias, C. 1998. Genetics of seabream (*Sparus aurata*). Study of microsatellites and their use for the estimation of genetic parameters of growth and other quantitative characters. Ph D. Thesis, University of Crete, Greece.
Bauer, MP, Goetz, FW. 2001. Isolation of gonadal mutations in adult zebrafish from a chemical mutagenesis screen. Biol Reprod, 64: 548–554.
Baumeister, SS, Kailasanath, V, Fernald, RD. 2007. Social dominance regulates androgen and estrogen receptor gene expression. Horm Behav, 51: 164–170.
Bender, N, Heg, D, Hamilton, IM, et al. 2006. The relationship between social status, behaviour, growth and steroids in male helpers and breeders of a cooperatively breeding cichlids. Horm Behav, 50: 173–182.
Bender, N, Heg-Bachar, Z, Oliveira, RF, et al. 2008. Hormonal control of brood care and social status in a cichlid fish with brood care. Physiol Behav, 94: 349–358.
Bengtsson, M, Hemberg, M, Rorsman, et al. 2008. Quantification of mRNA in single cells and modeling of RT-qPCR induced noise. BMC Mol Biol, 9: 63 doi:10.1186/1471-2199/9-63.
Bernet, D, Wahli, T, Kipfer, S, et al. 2009. Macroscopic gonadal deviations and intersex in developing whitefish *Coregonus lavaretus*.Aquat Biol, 6: 1–13.
Bernstein, VJ, Woldman, JR, Bemis, WE. 1997. Sturgeon Biodiversity and Conservation. Kluwer Academic Publishers, Dordrecht, The Netherlands.
Beulbens, K, Eding, EH, Gilson, P, et al. 1997. Gonadal differentiation intersexuality and sex ratios of European eel (*Anguilla anguilla* L.) maintained in captivity. Aquaculture, 153: 135–150.
Bisazza, A. 1997. Sexual selection constrained by internal fertilization in the live-bearing fish *Xenotoca eiseni*. Anim Beh, 54: 1347–1355.

Blacklidge, KH, Bidwell, OA. 1993. Three ploidy levels indicated by genome qualification in Acipenseriformes of North America. J Hered, 84: 427–430.
Blanc, JM, Poisson, H, Escaffre, AM, et al. 1993. Inheritance of fertilizing ability in male tetraploid rainbow trout (*Oncorhynchus mykiss*). Aquaculture, 110: 61–70.
Blazquez, M, Piferrer, F. 2004. Cloning, sequence analysis, tissue distribution and sex specific expression of the neural form of p450 aromatase in the juvenile sea bass (*Dicentrachus labrax*). Mol Cell Endocrinol, 219: 83–94.
Bobe, J, Guiguen, Y, Fostier, A. 2010. Diversity and biological significance of sex hormone-binding globulin in fish: an evolutionary perspective. Mol Cell Endocrinol, 316: 66–78.
Bogart, MH. 1987. Sex determination: A hypothesis based on steroid ratios. J Theor Biol, 128: 349–357.
Bohne, AS, Zhou, C, Froschauer, Q, et al. 2008. Identification of new gene candidates on sex chromosomes of the platyfish *Xiphophorus maculatus*. Cybium, 32: 69–71.
Bohne, A, Schultheis, C, Galiana-Arnoux, D, et al. 2009. Molecular analysis of the sex chromosomes of the platyfish *Xiphophorus maculatus*: Towards the identification of a new type of master sexual regulator in vertebrates. Integ Zool, 4: 272–284.
Bontems, F, Stein, A, Marlow, F, et al. 2009. Bucky ball organizes germplasm assembly in zebrafish. Curr Biol, 19: 414–422.
Booth, AJ, Buxton, CD. 1997. The biology of the panga *Pterogymnus laniarius* (Teleostei: Sparidae) on the Agulhas Bank, South Africa. Env Biol Fish, 49: 207–226.
Bowles, J, Schepers, G, Koopman, P. 2000. Phylogeny of the sox family of developmental transcription factors based on sequence and structural indicators. Dev Biol, 227: 239–255.
Braat, AR, Speksnijder, JE, Zivkovic, D. 1999. Germ line development in fishes. Int J Dev Biol, 43: 745–760.
Bradley, KM, Breyer, JP, Melville, DB, et al. 2011. An SNP based linkage map for zebrafish reveals sex determination loci. Genes/Genomes/Genetics, 1: doi 10.1534/93.111.000190.
Brantley, RK, Bass, AH. 1994. Alternative male spawning tactics and acoustic signals in the plainfin midshipman fish, *Porichthys notatus* Girard (Teleostei, Batrachoididae). Ethology, 96: 213–223.
Brennan, J, Capel, B. 2004. One tissue two fates: molecular genetic events that underlie testis versus ovary development. Nature Rev Genet, 5: 509–521.
Brouwer, SL, Griffiths, MH. 2005. Reproductive biology of carpenter seabream (*Argyrozona argyrozona*) (Pisces: Sparidae) in a marine protected area. Fish Bull, 103: 258–269.
Brum, MJI. 1996. Cytogenetic studies of Brazilian marine fish. Braz J Genet, 19: 421–427.
Brunelli, JP, Thorgaard, GH. 2004. A new Y chromosome specific marker for the Pacific salmon. Trans Am Fish Soc, 133: 10247–10253.
Brusle, S. 1988. Sex differentiation in teleosts: primordial germ cells as stem cells. Reproduction in fish: basic and applied aspects in endocrinology and genetics. Colloq Inst Natl Rech, Agron, 44: 21–24.
Brusle-Sicard, S, Reinboth, R. 1990. Protandric hermaphrodite peculiarities in *Amphiprion frenatus*. Brevoort (Teleostei: Pomacentridae). J Fish Biol, 36: 383–390.
Brusle-Sigard, S, Debas, L, Fourcault, B, et al. 1992. Ultrastructural study of sex inversion in a protogynous hermaphrodite *E. microdon* (Teleostei: Serranidae). Reprod Nat Dev, 32: 393–466.
Brusle-Sigard, S, Reinboth, R, Fourcault, B. 1994. Germinal potentialities during sexual state changes in a protandric hermaphrodite *Amphiprion frenatus* (Teleostei: Pomacentridae). J Fish Biol, 45: 597–611.
Bubley, WJ, Pashuk, O. 2010. Life history of a simultaneously hermaphroditic fish *Diplectrum formosum*. J Fish Biol, 77: 676–691.
Burns, JR, Weitzman, SH, Grier, HJ, et al. 1995. Internal fertilization, testis, sperm morphology in glanducaudine fishes (Teleostei: Characidae: Glandulocaudinae). J Morphol, 224: 131–145.

Butlin, R. 2002. Evolution of sex: the costs and benefits of new insights from old sexual lineage. Nat Rev Genet, 3: 311–317.

Buxton, CD. 1989. Protogynous hermaphroditism in *Chrysoblephus laticeps* (Cuvier) and *C. cristiceps* (Cuvier) (Teleostei: Sparidae). S Afr Tydsk Dierk, 24: 212–216.

Buxton, CD, Clarke, JR. 1992. The biology of the bronze bream *Pachymetopon grande* (Teleostei: Sparidae) from the south east Cape coast, South Africa. S Afr J Zool, 27: 21–32.

Cambell, WA, Yang, H, Zetterberg, H, et al. 2006. Zebrafish lacking Alzhmier presenilin enhancer 2 (pen-2) demonstrate excessive p53-dependent apoptosis and neuronal loss. J Neurochem, 96: 1423–1440.

Cao, J, Cao, Z, Wu, T. 2007. Generation of antibodies against *Dmrt1* and *Dmrt4* of *Oreochromis aureus* and analysis of their expression profile in *Oreochromis aureus* tissues. J Genet Genomics, 34: 497–509.

Cao, J, Chen, J, Wu, T, et al. 2010. Molecular cloning and sexually dimorphic expression of *DMRT4* gene in *Oreochromis aureus.* Mol Biol Rep, 37: 2781–2788.

Carrell, DT. 2011. Epigenetic marks in zebrafish sperm: insights into chromatin compaction, maintenance of pluripotency and the role of the paternal genome after fertilization. Asian J Biotechnol, 13: 620–621.

Carruth, LL. 2000. Freshwater cichlid *Crenicara punctulata* is a protogynous sequential hermaphrodite. Copeia, 2000: 71–82.

Carvalho, AB. 2003. The advantages of recombination. Nat Genet, 34: 128–129.

Cek, S, Bromage, N, Randall, C, et al. 1998. The reproductive biology of the rosy barb (*Puntius conchonius*): Sex differentiation and gametogenesis. pp167–176. In: MS Celikkale, E Duzgunes, I Okumus, C Mutlu (eds) Proc First Internatl Symp Fish Ecol. Trabzon, Turkey.

Chang, CF, Lau, EL, Lin, BY. 1995. Stimulation of spermatogenesis or of sex change according to the doses of exogenous estradiol-17β in juvenile males of protandrous black porgy *Acanthopagrus schlegeli.* Gen Comp Endocroniol, 100: 355–367.

Chang, CF, Lin, BY, Lau, EL, et al. 1997. The endocrine mechanism of sex reversal in the protandrous black porgy *Acanthopagrus schlegeli*: a review. Chin J Physiol, 40: 197–205.

Chang, X, Kobayashi, T, Senthilkumaran, B, et al. 2005. Two types of aromatase with different encoding genes, tissue distribution and developmental expression in Nile tilapia (*Oreochromis nilotcus*). Gen Comp Endocrinol, 141: 101–115.

Chaves-Pozo, E, Miulero, V, Meseguer, J, et al. 2005. An overview of cell renewal in the testis throughout the reproductive cycle of a seasonal breeding teleost, the gilthead seabream (*Sparus aurata* L.). Biol Reprod, 72: 593–601.

Chen, J, Yuan, H, Sun, D, et al. 2006. Sequence and expression of three members of the *Sox* gene in Amur sturgeon (*Acipenser schrenckii*). J Appl Ichthyol, 22 (Suppl. 1): 77–81.

Cherr, GN, Clark, WH. 1985. Gamete interaction in the white sturgeon *Acipenser transmontanus*: a morphological and physiological review. Env Biol Fish, 14: 11–22.

Cheshenko, K, Pakdel, F, Segner, H, et al. 2008. Interference of endocrine chemicals with aromatase *cyp19* expression or activity and consequences for reproduction of teleost fish. Gen Comp Endocrinol, 155: 31–62.

Chiang, EF-L, Yan, Y-L, Guiguen, Y, et al. 2001. Two *cyp19* (p450 aromatase) genes on duplicated zebrafish chromosomes are expressed in ovary or brain. J Mol Evol, 18: 542–550.

Choi, I, Oh, J, Cho, BN, et al. 2004. Characterization and comparative genomic analysis of intronless Adams with testicular expression. Genomics, 83: 636–646.

Chopelet, J, Waples, RS, Mariani, S. 2009. Sex change and the genetic structure of marine fish. Fish Fish, 10: 329–343.

Chu-Koo, F, Dugue, R, Aguilar, MA, et al. 2009. Gender determination in the paiche or pirarucu (*Arapaima gigas*) using plasma vitellogenin, 17β estradiol and 11 Ketotestosterone. Fish Physiol Biochem, 35: 125–136.

Coetzee, PS. 1983. Seasonal histological and macroscopic changes in the gonads of *Cheimerius nufar* (Ehrenberg, 1820) (Sparidae: Pisces). S Afr J Zool, 18: 76–88.

Coetzee, PS. 1986. Diet composition and breeding cycle of blacktail, *Diplodus sargus capensis* (Pisces: Sparidae), caught off St. Croix Island, Algoa Bay, South Africa. S Afr J Zool, 21: 237–243.

Cole, KS. 2002. Gonad morphology, sexual development and colony composition in the obligate coral-dwelling damselfish *Dascyllus aruanus*. Mar Biol, 140: 151–163.

Cole, KS. 2008. Transient ontogenetic expression of hermaphroditic gonad morphology within the Gobiosoma group of the Neotropical seven-spined gobies (Teleostei: Gobiidae). Mar Biol, 154: 943–951.

Cole, KS, Shapiro, DY. 1992. Gonad structure and hermaphroditism in the gobiid genus *Coryphopterus* (Teleostei: Gobiidae). Copeia, 1992: 996–1003.

Cole, KS, Hoese, DF. 2001. Gonad morphology, colony demography and evidence for hermaphroditism in *Gobiodon okinawae* (Teleostei: Gobiidae). Env Biol Fish, 61: 161–173.

Comfort, A. 1961. Age and reproduction in female *Lebistes*. Gerontologia, 5: 146–149.

Coons, A, Creech HJ, Jones, RN. 1941. Immunological properties of an antibody containing a fluorescent group. Proc Soc Exp Biol Med, 47: 200–202.

Coughlan, T, Schartl, M, Hornung, U, et al. 1999. PCR-based sex test for *Xiphorphorus maculatus*. J Fish Biol, 54: 218–222.

Coward, K, Bromage, NR, 1999. Spawning periodicity, fecundity and egg size in laboratory held stocks of a subtrate-spawning tilapiine *Tilapia zilli*. Aquaculture, 171: 251–267 .

Crespi, BJ, Teo, R. 2002. Comparative phylogenetic analysis of the evolution of semelparity and life history in salmonid fishes. Evolution, 56: 1008–1020.

Csenki, Z, Zaucker, A, Kovacs, B, et al. 2010. Intraovarian transplantation of stage I-II follicles results in viable zebrafish embryos. Int J Dev Biol, 54: 585–589.

Cui, J-Z, Shen, X-Y, Gong, Q-L, et al. 2006. Identification of sex markers by $_{C}$DNA-AFLP in *Takifugu rubripes*. Aquaculture, 257: 30–36.

David, CJ. 2004. Experimental sperm preservation and genetic studies in selected fish. PhD. Thesis, Madurai Kamaraj University, Madurai, India.

David, CJ, Pandian, TJ. 2006. Cadaveric sperm induces intergeneric androgenesis in the fish *Hemigrammus caudovittatus*. Theriogenology, 65: 1048–1070.

Davidson, WS, Huang, T-K, Fujiki, K, et al. 2009. The sex determining loci and sex chromosomes in the family Salmonidae. Sex Dev, 3: 78–87.

de Bruin, J-P, Gosden, RG, Finch, CE, et al. 2004. Ovarian aging in two species of long lived rockfish *Sebastes aleutianus* and *S. alutus*. Biol Reprod, 71: 1036–1042.

Delvin, RH, Nagahama, Y. 2002. Sex determination and sex differentiation in fish: an overview of genetic, physiological and environmental influences. Aquaculture, 208: 191–364.

Devlin, RH, Biagi, CA, Smailus, DE. 2001. Genetic mapping of Y-chromosomal DNA markers in Pacific salmon. Genetica, 111: 43–58.

Deng, X, Deng, ZL, Cai, MY. 1991. Spawning population characteristics of *Acipsenser sinensis* in Yangtze River just below Gezhouba Dam. In: P Willior (ed). Acipenser. CEMAGREF Publ, Bordeaux, France.

Dieckmann, U, Heino, M. 2007. Probabalistic maturation reaction norms: their history, strengths and limitations. Mar Eiol Prog Ser, 335: 253–269.

Ding, L, Kuhne, WW, Hinton, DE, et al. 2010. Quantifiable biomarkers of normal aging in the Japanese medaka fish (*Oryzias latipes*). PLoS ONE, 5: e13287. doi:10.1371/journal.pone.0013287.

Dorsch, R, Wagner, DS, Mintzer, KA, et al. 2004. Maternal control of vertebrate development before the midblastula transition: Mutants from the zebrafish. T Dev Cell, 6: 771–780.

Douard, V, Brunet, F, Boussau, B, et al. 2008. The fate of the duplicated androgen receptor in fishes: a late neo-functionalization event? BMC Evol Biol, 8: doi:10.1186/1471-2148-8-336.

Draper, BW, McCallum, CM, Moens, CB. 2007. *nanos* 1 is required to maintain oocyte production in adult zebrafish. Dev Biol, 305: 589–598.

Du, SJ, Devlin, RH, Hew, CL. 1993. Genomic structure of growth hormone genes in chinook salmon (*Oncorhynchus tshawytscha*): presence of two functional genes GH-I and GH-II and a male specific pseudogene GH-ψ. Cell Biol, 12: 739–751.

Du, Q-Y, Wang, F-Y, Hua, H-Y, et al 2007. Cloning and study of adult tissue specific expression of *Sox9* in *Cyprinus carpio*. J Genet, 86: 85–91.

Du, XZ, Zhou, L, Zhao, HB, et al. 2008. Identical sequences but different expression patterns of *Hira* gene in gynogenetic and gonochoristic crucian carps. Fish Physiol Biochem, 34: 175–184.

Duponchelle, F, Cecchi, P, Corbin, D, et al. 2000. Variations in fecundity and egg size of female Nile tilapia *Oxochromis niloticus* from man-made lakes of Cote d'Ivoire. Env Fish Biol, 57: 155–170.

Ebisawa, A. 1997. Some aspects of reproduction and sexuality in the spotcheek emperor, *Lethrinus rubrioperculatus,* in waters off the Ryukyu Islands. Ichthyol Res, 44: 201–212.

Engelhard, GH, Henio, M. 2004. Maturity changes in Norwegian spring-spawning herring *Clupea harengus:* compensatory or evolutionary responses? Mar Ecol Prog Ser, 272: 245–256.

Erickson, DL, Webb, MAH. 2007. Spawning periodicity, spawning migration and size at maturity of green sturgeon *Acipenser mediorostris* in the Rough River, Oregon. Env Biol Fish, 79: 255–268.

Erisman, BE, Rosales-Casian, JA, Hastings, PA. 2008. Evidence of gonochorism in a grouper *Mycteroperca rosacea* from the Gulf of California, Mexico. Env Biol Fish, 82: 23–33.

Farlora, R, Kobayashi, S, Franca, LR, et al. 2009. Expression of GFP in transgenic tilapia under the control of the medaka β-actin promoter: establishment of a model system for germ cell transplantation. Anim Reprod, 6: 450–459.

Feistma, H, Leal, MC, Moens, PB, et al. 2007. Mlh1 deficiency in zebrafish results in male sterility and aneuploid as well as triploid progeny in females. Genetics, 175: 1561–1569.

Fenske, M, Segner, H. 2004. Aromatase modulation alters gonadal differentiation in developing zebrafish (*Danio rerio*). Aquat Toxicol, 67: 105–126.

Fernandino, JI, Hattori, RS, Kimura, H, et al. 2008. Expression profile and estrogenic regulation of anti-Mullerian hormone during gonadal development in pejerrey *Odentesthes bonariensis*, a teleost fish with strong temperature-dependent sex determination. Dev Dyn 237: 3192–3199.

Fernandino, JI, Popesxu, JT, Paul-Prasanth, B, et al. 2010. Analysis of sexually dimorphic expression of genes at early gonadogenesis of pejerrey *Odontesthes bonariensis* using heterologous micro-array. Sex Dev, DOI:10.1159/000324423.

Filby, AL, Tyler, CR. 2005. Molecular characterization of estrogen profiles in fathead minnow (*Pimephales promelas*). Biol Reprod, 73: 121–124.

Fishelson, L. 1992. Comparative gonad morphology and sexuality of the Muraenidae (Pisces: Teleostei). Copeia, 1992: 197 -209.

Fitzpatrick, JL, Desjardins, JK, Milligan, N, et al. 2007. Reproductive-tactic-specific variation in sperm swimming speeds in a shell-brooding cichlid. Biol Reprod, 77: 280–284.

Foran, CM, Bass, AH. 1999. Preoptic GnRH and AVT: axes for sexual plasticity in teleost fish. Gen Comp Endocinol, 116: 141–152.

Frisch, AJ, Walker, SPW, McCormick, MI, et al. 2007. Regulation of protogynous sex change by competition between corticosteroids and androgens: An experimental test using sandperch *Parapercis cylindrica.* Horm Behav, 52: 540–545.

Fuji, K, Yoshida, K, Hattori, K, et al. 2010. Identification of the sex-linked locus in yellowtail *Seriola quinqueradiata.* Aquaculture, 308: 551–555.

Fukamachi, S, Kinoshita, M, Aizawa, K, et al. 2009. Dual control by a single gene of secondary sexual characters and mating preferences in medaka. BMC Biol, 7: 64 doi:10.1186/1741-7007-7- 64.

Furutani-Seiki, M, Sasado, T, Morinaga, C, et al. 2004. A systematic genome-wide screen for mutations affecting organogenesis in medaka *Oryzias latipes.* Mech Dev, 121: 647–658.

Gale, ML, Fitzpatrick, MS, Lucero, M, et al. 1999. Masculinization of Nile tilapia (*Oreochromis niloticus*) by immersion in androgens. Aquaculture, 178: 349–357.
Gao, Z, Wang, H-P, Rapp, D, et al. 2009. Gonadal sex differentiation in the bluegill *Lepomis macrochirus* and its relation to fish size and age. Aquaculture, 294: 138–146.
Garcia-Diaz, MM, Lorente, MJ, Gonzalez, JA, et al. 2002. Morphology of the ovotestis of *Serranus atricauda* (Teleostei: Serranidae). Aquat Sci, 64: 87–96.
Garcia-Lopez, A, Jonge H de, Nobrega, RH, et al. 2010. Studies in zebrafish reveal unusual cellular expression patterns of gonadotropin receptor messenger ribonucleic acid in the testis and unexpected functional differentiation of the gonadotropins. Endocrinology, 151: 2349–2360.
Gardner, L, Anderson, T, Place, AR, et al. 2003. Sex change strategy and the aromatase. Fish Physiol Biochem, 28: 147–148.
Gavaia, PJ, Dinis, MT, Cancela, ML. 2002. Osteological development and abnormalities of the vertebral column and caudal skeleton in larval and juvenile stages of hatchery-reared Senegal sole (*Solea senegalensis*). Aquaculture, 211: 305–323.
George, T, Pandian, TJ. 1995. Production of ZZ females in the female-heterogametic black molly *Poecilia sphenops* by endocrine sex reversal and progeny testing. Aquaculture, 136: 81–90.
George, T, Pandian, TJ. 1997. Interspecific hybridization in poeciliids. Indian J Exp Biol, 35: 628–637.
George, T, Pandian, TJ. 1998. Dietary administration of androgens induces sterility in the female heterogametic black molly *Poecilia sphenops* (Cvvier & Valenciennes, 1846). Aquacult Res, 29: 169–175.
Gerhard, GS, Kauffman, EJ, Wang, X, et al. 2002. Life spans and senescent phenotypes in two strains of zebrafish (*Danio rerio*). Exp Gerontol, 37: 1055–1068.
Gnaani, A, Lee, B-7, Zilberman, N, et al. 2008. Genetics of sex determination in tilapia species. Sex Dev, 2: 43–54.
Goetz, FW, MacKenzie, S. 2008. Functional genomics with microarrays in fish biology and fisheries. Fish Fish, 9: 378–395.
Goncalves, D, Telejoao, M, Rui, A, et al. 2008. Brain and gonadal aromatase activity and steroid hormone levels in polymorphic males of the peacock blenny *Salaria pavo*. Horm Behav, 54: 717–725.
Goodwin, NB, Dulvy, NK, Reynolds, JD. 2002. Life history correlates of the evolution of live bearing in fishes. Phil Trans R Soc, Lond, 357B: 259–267.
Gorman, KF, Breden, F. 2007. Teleosts as models for human vertebral stability and deformity. Comp Biochem Physiol, 145C: 28–38.
Gorman, KF, Tredwell, SJ, Breden, F. 2007. The mutant guppy syndrome *Curveback* as a model for human heritable spinal curvature. Spine, 32: 735–741.
Goswami, SV, Sundararaj, BI. 1968. Compensatory hypertrophy of the remaining ovary after unilateral ovariectomy at various phases of reproductive cycle of catfish *Heteropneustes fossilis* Bloch. Gen Comp Endocrinol, 11: 401–413.
Grandi, G, Colombo, G. 1997. Development and early differentiation of gonad in the European eel (*Anguilla anguilla* [L.] Anguilliformes, Teleostei): A cytological and ultrastructural study. J Morphol, 231: 195–216.
Grandi, G, Chicca, M, Tagliati, E, et al. 2010. Immunolocalization of steroidogenic enzymes in gonads of European eel, *Anguilla anguilla* (L.). J Appl Ichthyol, 26: 40–48.
Graves, JAM, Peichel, CL. 2010. Are homologies in vertebrate sex determination due to shared ancestry or to limited options? Genome Biol, 11: 205, doi:10.1186/gb-2010-11-4-205.
Greytak, SR, Callard, GV. 2007. Cloning of three estrogen receptors (ER) from killifish (*Fundulus heteroclitus*): Differences in populations from polluted and reference environments. Gen Comp Endocrinol, 150: 174–178.
Gross, MR. 1982. Sneakers, satellites and parentals: Polymorphic alternative strategies in North American sunfishes. Z Tierpsychol, 60: 1–26.

Gross, MR. 1984. Sunfish, salmon and evolution of alternative reproductive strategies and tactics in fishes. pp 55–75. In: GW Potts, RJ Wooton (eds). Fish Reproduction: Strategies and Tactics. Acadamic Press, London, UK.

Guan, G, Kobayashi, T, Nagahama, Y. 2000. Sexually dimorphic expression of two types of *DM* (Double sex/Mab-3)-domain genes in a teleost, the tilapia (*Oreochromis niloticus*). Biochem Biophys Res Commun, 272: 662–666.

Guiguen, Y, Baroiller, JR, Ricordel, MJ, et al. 1999. Involvement of estrogen in the process of sex differentiation in two fish species ; the rainbow trout (*Oncorhynchus mykiss*) and tilapia (*Oreochromis niloticus*). Mol Reprod Dev, 54: 154–162.

Guiguen, Y, Fostier, A, Piferrer, F, et al. 2010. Ovarian aromatase and estrogens: A pivotal role for gonadal sex differentiation and sex change in fish. Gen Comp Endocrinol, 165: 352–366.

Gunter, HM, Clabaut, C, Salzburger, W, et al. 2011. Identification and characterization of gene expression involved in the coloration of cichlid fish using microarray and qRT-PCR approaches. J Mol Evol, 72: 127–137.

Guo, Y, Li, Q, Gao, S, et al. 2004. Molecular cloning, characterization, and expression in brain and gonad of Dmrt5 of zebrafish. Biochem Biophys Res Commun, 324: 569–575.

Guo, Y, Cheng, H, Huang, X, et al. 2005. Gene structure, multiple alternative splicing and expression in gonads of zebrafish (Dmrt1). Biochem Biophys Res Commun, 330: 950–957.

Guo, B-C, JunBing, LI, Tong, CB, et al. 2008a. Cloning and sequence analysis of *sox* genes in a tetraploid fish *Tor douronensis*. Chinse Sci Bull, 53: 1988–1995.

Guo, X, Yan, J, Liu, S, et al. 2008b. Isolation and expression analysis of Sox9a gene in triploid crucian carp. Fish Physiol Biochem, Doi, 10: 1007/s 10695-008-92092.

Guraya, SS. 2000. The biology of gonadal development, sex differentiation and maturation and sex reversal in fish: Cellular, molecular and endocrinological aspects. Proc Indian Natl Acad Sci 66B: 167–194.

Hale, MC, Jackson, JR, DeWoody JA. 2010. Discovery and evaluation of candidate sex-determining genes and xenobiotics in the gonads of lake sturgeon (*Acipenser fulvescens*). Genetica, DO1 10.1007/s 10709-010-9455-y.

Halm, SG, Mortinez-Rodriquez, L, Rodriquez, F, et al. 2004. Cloning, characterization and expression of three estrogen receptors (ER alpha, ER beta2) in the European sea bass *Dicentrarchus labrax*. Mol Cell Endocrinol, 223: 63–75.

Hamaguchi, S. 1982. A light and electron microscopic study on the migration of primordial germ cells in the teleost *Oryzias latipes*. Cell Tissue Res, 227: 139–151.

Hamaguchi, S, Sakaizumi, M. 1992. Sexually differentiated mechanisms of sterility in interspecific hybrids between *Oryzias latipes* and *O. curvinotus*. J Exp Zool, 263: 323–329.

Hansen, OJ, Puvanendran, V. 2010. Fertilization success and blastomere morphology as predictors of egg and juvenile quality of domesticated *Gadus morhua* broodstock. Aquacult Res, 41: 1791–1798.

Hannenhalli, S, Kaestner, KH. 2009. The evolution of Fox genes and their role in development and disease. Nature Rev Genet, 10: 233–240.

Harbott, LK, Baumeister, SS, White, RB, et al. 2007. Androgen receptors in a cichlid fish *Astatotilapia burtoni:* structure, localization and expression levels. J Comp Neur, 504: 57–73.

Harrington, RW Jr. 1967. Environmentally controlled induction of primary male gonochorists from the eggs self fertilizing hermaphroditic fish *Rivulus marmoratus* Poey. Biol Bull, 132: 174–199.

Harrington, RW Jr. 1975. Sex determination and differentiation among uniparental homozygotes of the hermaphroditic fish *Rivulus marmoratus* (Cyprinodontidae: Atherniformes). In: R Reinboth, (ed) Intersexuality in the Animal Kingdom. Springer Verlag, Berlin, Germany.

Haukenes, AH, Plante, S, Buck, CL. 2008. Pattern of sex steroid concentrations in relation to the annual reproductive cycle of wild and captive black rockfish. North Am J Aquacult, 70: 212–218.

Hawkins, MB, Thomas, P. 2004. The unusual binding properties of the third distinct teleost estrogen receptor subtype ER beta are accompanied by highly conserved amino acid changes in the ligand binding domain. Endocrinology, 145: 2968–2977.

Hawkins, MB, Thornton, JW, Crews, D, et al. 2000. Identification of a third distinct estrogen receptor and reclassification of estrogen receptors in teleosts. Proc Natl Acad Sci USA, 97: 10751–10756.

He, CL, Du, JL, Wu, GC, et al. 2003. Differential Dmrt1 transcripts in gonads of the protandrous black porgy *Acanthopagrus schlegeli*. Cytogenet Genome Res, 101: 309–313.

Heath, DD, Rankin, L, Bryden, CA, et al. 2002. Heritability and Y-chromosome influence in the jack male life history of chinook salmon (*Oncorhynchus tshawystcha*). Heredity, 89: 311–317.

Hellqvist, A, Marjer, I, Borg, B. 2002. Effects of hemi-castratuion on plasma steroid levels in two teleost fishes; the three-spined stickleback *Gasterosteus aculeatus* and the Atlantic salmon *Salmo salar*. Fish Physiol Biochem, 26: 107–110.

Herpin, A, Schartl, M. 2011. Dmrt1 genes at the crossroads: a widespread and central class of sexual development factors in fish. FEBS J, 278: 1010–119.

Herpin, A, Schindler, D, Kraiss, A, et al. 2007. Inhibition of primordial germ cell proliferation by the medaka male determining *Dmrt1bY*. BMC Dev Biol, 7: 99, doi: 10.1186/1471-213X-7/99.

Herpin, A, Braasch, I, Kraeussling, M, et al. 2010. Transcriptional rewiring of the sex determining *dmrt1*gene duplicate by transposable elements. PLoS Genet, 6: e1000844. doi:10.1371/journal. Gen.1000844.

Hett, AK, Pitra, C, Jenneckens, I, et al. 2005. Characterization of Sox9 in European Atlantic sturgeon (*Acipenser sturio*). J Hered, 96: 150–154.

Higa, M, Kanda, H, Kitahashi, T, et al. 2000. Quantitative analysis of *fushi taraza* factor 1 homolog messenger ribonuclic acids in the pituitary of salmon at different pre-spawning stages. Biol Reprod, 63: 1756–1763.

Higaki, S, Eto, Y, Kawakami, Y, et al. 2010. Production of fertile zebrafish (*Danio rerio*) possessing germ cells (gametes) originated from primordial germ cells recovered from vitrified embryos. Reproduction, 139: 733–740.

Higuchi, R, Dollinger, G, Walsh, PS, et al. 1992. Simultaneous amplification and detection of specific DNA sequences. Biotechnology, 10: 413–417.

Hobby, AC, Pankhurst, NW, Geraphty, DP, 2000. A comparison of sex steroid binding protein (SBP) in four species of teleost fish. Fish Physiol Biochem, 23: 245–250.

Holland, MC, Hassin, S, Zohar, Y. 2000. Gonadal development and plasma steroid levels during pubertal development in captive-reared striped bass *Morone saxatilis*. J Exp Zool, 286: 49–63.

Hong, SK, Levin, CS, Brown, J, et al. 2010. Pre-gastrula expression of zebrafish extra-embryonic genes. Dev Biol, 10: 42,doi:10.1186/1471-213x-10-42.

Hong, Y, Winkler, C, Schartl, M. 1996. Pluripotency and differentiation of embryonic stem cell lines from the medaka fish (*Oryzias latipes*). Mech Dev, 60: 33–44.

Hong, Y, Winkler, C, Schartl, M. 1998. Production of medaka fish chimeras from a stable embryonic stem cell line. Proc Natl Acad Sci USA, 95: 3679–3684.

Hong, Y, Liu, T, Zhao, H, et al. 2004. Establishment of a normal spermatogonial germ cell line capable of sperm production *in vitro*. Proc Nal Acad Sci USA, 101: 8011–8016.

Horth, L, Binckley, C, Wilk, R, et al. 2010. Color, body size and genitalia size are correlated traits in eastern mosquitofish (*Gambusia holbrooki*). Copeia, 2010: 196–202.

Houwing, S, Kamminga, LM, Berezikov, E, et al. 2007. A role for Piwi and piRNAs in germ cell maintenance and transposon silencing in zebrafish. Cell, 129: 69–82.

Hsiao, CD, Tsai, HJ. 2003. Transgenic zebrafish with fluorescent germ cell: a useful tool to visualize germ cell proliferation and juvenile hermaphroditism *in vivo*. Dev Biol, 262: 313–323.

Hsu, C-C, Hou, M-F, Hong, J-R, et al. 2010. Inducible male sterility by targeted cell ablation in zebrafish testis. Mar Biotechnol, 12: 466–478.

Huang, B, Wang, S, Ning, Y, et al. 1999. Autosomal XX sex reversal caused by duplication of sox9. Am J Med Genet, 87: 349–353.

Huang, X, Guo, Y, Shui, Y, et al. 2005. Multiple alternative splicing and differential expression of *dmrt1* during gonad transformation of rice field eel. Biol Reprod, 73: 1017–1024.

Hunter, GA, Donaldson, EM, 1983. Hormonal sex control and its application to fish culture. pp 223–303. In: WS Hoar, DJ Randall, EM Donaldson (eds). Fish Physiology. Academic Press, New York, USA.

Ijiri, S, Kaneko, H, Kobayashi, T, et al. 2008. Sexual dimorphic expression of genes in gonads during early differentiation of teleost fish, the Nile tilapia *Oreochromis niloticus*. Biol Reprod, 78: 333–341.

Ikeuchi, T, Todo, T, Kobayashi, T, et al. 2001. Two subtypes of androgen and progesterone receptors in fish testes. Comp Biochem Physiol B: Biochem Mol Biol, 129: 449–455.

Jamieson, BGM. 1991. Fish Evolution and Systematic Evidence from Spermatozoa. Cambridge University Press, Cambridge, UK.

Jeong, HB, Park, JG, Park, YJ, et al. 2009. Isolation and characterization of DMRT1 and its putative regulatory region in the protogynous wrasse *Halichoeres tenuispinis*. Gene 438: 8–16.

Jiang, JQ, Kobayashi, T, Ge, W, et al. 1996. Fish testicular 11β hydroxylase: cDNA cloning and mRNA expression during spermatogenesis. FEBS Lett, 397: 250–252.

Jiang, Y, Li, WS, Xie, J, et al. 2003. Sequence and expression of a cDNA encoding both pituitary adenylate cyclase activating polypeptide and growth hormone-releasing hormone in grouper (*Epinephelus coioides*). Acta Biochim Biophys Sin, 35: 864–872.

Jo, PG, An, KW, Kim, NN, et al. 2007. Induced expression of *doublesex-and mab-3-related transcription factor 1* (DMRT-1) mRNA by testosterone in the olive flounder *Paralichthys olivaceus* ovary. J Aquac, 20: 199–202.

Johnson, H, Seppola, M, Torgersen, JS, et al. 2010. Sexually dimorphic expression of dmrt1 in immature and mature Atlantic cod (*Gadus morhua* L.). Comp Biochem Physiol, 156B: 197–205.

Johnson, J, Canning, J, Kaneko, T, et al. 2004. Germline stem cells and follicular renewal in the postnatal mammalian ovary. Nature, 428: 145–149.

Jorgensen, A, Morthorst, JE, Andersen, O, et al. 2008. Expression profiles of six zebrafish genes during gonadal sex differentiation. Reprod Biol Endocrinol, 6:25 doi:10.1186/1477-7827-6-25.

Joubert, CSW. 1981. Aspects of the biology of five species of inshore reef fishes on the Natal coast, South Africa. Invest Rep Occanogr Inst, 51: 1–16.

Kaczarczyk, AN, Kopp, A. 2010. Germline stem cell maintenance as a proximate mechanism of life-history trade off ?. Bioessays: DOI 10.1002/bies.201000085.

Kallivretaki, E, Eggen, RIL, Neuhauss, SCF, et al. 2007. The zebrafish brain specific aromatase *cyp19a2* is neither expressed nor distributed in a sexually dimorphic manner during sexual differentiation. Dev Mech, 236: 3155–3166.

Kallman, KD. 1984. A new look at sex determination in poeciliid fishes pp 97–171. In: BJ Turner (ed). Evolutionary Genetics of Fishes. Plenum Publishing, New York, USA.

Katano, O. 1983. Spawning of the dark chub *Zucco temmincki* (Temmincket Schlegel). (Pisces: Cyprinidae) with reference to male social status. Phy Biol Ecol Japan, 20: 33–52.

Katano, O. 1992. Spawning tactics of paired males of the dark chub. *Zucco temmincki* reflect potential fitness costs of satellites. Env Biol Fish, 35: 343–350.

Kavumpurath, S, Pandian, TJ. 1992. Production of YY male in the guppy *Poecilia reticulata* by endocrine sex reversal and progeny testing. Asian Fish Sci, 5: 265–276.

Kavumpurath, S, Pandian, TJ. 1993. Masculinization of *Poecilia reticulata* by dietary administration of synthetic or natural androgen to gravid females. Aquaculture, 116: 83–89.

Kazancioglu, E, Alonzo, SH. 2009. Costs of changing sex do not explain why sequential hermaphroditism is rare. Am Nat, 173: 327–336.

Kazianis, S, Nairn, RS, Walter, RB, et al. 2004. The genetic map of *Xiphophorus* fishes represented by 24 multipoint linkage groups. Zebrafish, 1: 287–304.

Kersten, CA, Krisfalusi, M, Parsons, JE, et al. 2001. Gonadal regeneration in masculinized female or steroid treated rainbow trout (*Oncorhynchus mykiss*). J Exp Zool, 290: 396–401.

Khoo, L, Lim, TM, Chan, WK, et al. 1999a. Linkage analysis of three sex-linked colour pattern genes in the guppy *Poecilia reticulata*. Zool Sci, 16: 893–903.

Khoo, L, Lim, TM, Chan, WK, et al. 1999b. Sex linkage on the black tail caudal peduncle and red tail genes in the tuxedo strain of the guppy *Poecilia reticulata*. Mar Biotechnol, 5: 273–293.

Kikuchi, K, Kai, W, Hosokawa, A, et al. 2007. The sex determining locus in the tiger pufferfish *Takifugu rubripes*. Genetics, 175: 2039–2042.

Kim, S, Osakawa, JK, Park, JG, et al. 2002. Sequence and expression of androgen receptor and estrogen receptor gene in the sex types of protogynous wrasse *Halichoeres trimaculatus*. Gen Comp Endocrinol, 127: 165–173.

Kimmel, CB, Ullman, B, Walker, C, et al. 2005. Evolution and development of facial bone morphology in three spine stickleback. Proc Natl Acad Sci USA, 102: 5791–5796.

King, ML, Zhou, Y, Bubunenko, M. 1999. Polarizing genetic information in the egg: RNA localization in the frog oocyte. Bioessays, 21: 546–557.

Kirankumar, S, Pandian, TJ. 2003. Effect on growth and reproduction of hormone immersed and masculinized fighting fish *Betta splendens*. J Exp Zool, 293: 606–616.

Kirankumar, S, Anathy, V, Pandian, TJ. 2003. Hormonal induction of supermale golden rosy barb and isolation of Y chromosome specific markers. Gen Comp Endocrinol, 134: 62–71.

Kirpichnikov, VS. 1981. Genetic Basis of Fish Selection. Springer Verlag, Berlin, Germany.

Kishi, S, Uchiyama, J, Baughman, AM, et al. 2003. The zebrafish as a vertebrate model of functional aging and very gradual senescence. Exp Gerontol, 38: 777–786.

Kishi, S, Bayliss, PE, Uchiyama, J, et al. 2008. The identification of zebrafish mutants showing alterations in senescence-associated biomarkers. PLoS Genet, 4: e1000152. doi:10.1371/journal.pgen.1000152.

Kitano, T, Takemune, K, Nagahama, Y, et al. 2000. Aromatase inhibitor and 17-α methytestosterone cause sex reversal from genetic female to phenotypic males and suppression of p450 aromatase gene expression in Japanese flounder (*Paralichthys olivaceus*). Mol Reprod Dev, 56: 1–5.

Kitano, T, Koyanagai, T, Adachi, R, et al. 2006. Assessment of estrogenic chemicals using an estrogen receptor gene assay in fish. Mar Biol, 149: 49–55.

Kitano, T, Yoshinaga, N, Shiraishi, E, et al. 2007. Tamoxifen induces masculinization of genetic females and regulates p450 aromatase and Mullerian inhibiting substance mRNA expression in Japanese flounder (*Paralichthys olivaceus*). Mol Reprod Dev, 74: 1171–1177.

Kline, RJ, Khan, IA, Holt, GJ. 2011. Behavior, color change and time for sexual inversion in the protogynous grouper (*Epinephelus adscensionis*). PLoS ONE, 6: e19576. doi:10.1371/journal. pone.0019576.

Kluver, N, Herpin, A, Braasch, I, et al. 2009. Regulatory back-up circuit of medaka Wt1 co-orthologs ensures PGC maintenance. Dev Biol, 325: 179–188.

Knapp, R. 2004. Endocrine mediation of vertebrate male alternate reproductive tactics: the next generation of studies. Integ Comp Biol, 43: 658–668.

Knaut, H, Pelegri, F, Bohmann, K, et al. 2000. Zebrafish *vasa* RNA but not its protein component of the germplasm and segregates asymmetrically before germline specification. J Cell Biol, 149: 875–888.

Knaut, H, Steinbeisser, H, Schwarz, H, et al. 2002. An evolutionary conserved region of the *vasa* 3[1] UTR targets RNA translation to the germ cells in teleost fish *Oreochromis niloticus*. Mech Dev, 99: 139–142.
Kobayashi, K. 1999. Intersexuality and sexual function in some Japanese wrasses (Teleostei, Labridae). Sci Rep Mus, Tokai Univ, 1: 37–88.
Kobayashi, K, Suzuki, K. 1994. Intersexuality and reproduction in *Halichoeres poecilopterus* in Suruga Bay, Central Japan. J Fac Mar Sci Technol, Tokai Univ, 38: 233–256.
Kobayashi, K, Suzuki, Y, Shiobara, Y. 1993. Reproduction and hermaphroditism in *Parapercis snyderi* (Teleosteic: Parapercidae) in Suruga Bay, Central Japan. J Fac Mar Sci Technol, Tokai Univ, 35: 149–168.
Kobayashi, M, Aida, K, Stacey, NE. 1991. Induction of testis development by implantation of 11-Ketotestosterone in female goldfish. Zool Sci, 8: 389–393.
Kobayashi, T. 2010. *In vitro* germ cell differentiation during sex differentiation in a teleost fish. Int J Dev Biol, 54: 105–111.
Kobayashi, T, Kajiura-Kobayashi, H, Nagahama, Y. 2000. Differential expression of *vasa* homologue gene in the germ cells during oogenesis and spermatogenesis in a teleost fish *Oreochromis niloticus*. Mech Dev, 99: 139–142.
Kobayashi, T, Kajiura-Kobayashi, H, Nagahama, Y. 2003. Induction of XY sex reversal by estrogen involves altered gene expression in a teleost. Cytogenet Genome Res, 101: 289–294.
Kobayashi, T, Matsuda, M, Kajiura-Kobayashi, H, et al. 2004. Two DM-domain genes *DMY* and DMRT1 involved in testicular differentiation and development in medaka *Oryzias latipes*. Dev Dyn, 231: 518–526.
Kobayashi, T, Kajiura-Kobayashi, H, Guan, G, et al. 2008. Sexual dimorphic expression of DMRT and Sox9a during gonadal differentiation and hormone-induced sex reversal in the teleost fish Nile tilapia (*Oreochromis niloticus*) Dev Dyn 237: 297–306.
Kobayashi, T, Ishibashi, R, Yamamoto, S, et al. 2011. Gonadal morphogenesis and sex differentiation in cultured chub mackerel *Scomber japonicus*. Aquacult Res, 42: 230–239.
Kobayashi, Y, Nakamura, M, Sunobe, T, et al. 2009. Sex change in the gobiid fish is mediated through rapid switching of gonadotropin receptors from ovarian to testicular portion or vice versa. Endocrinology, 150: 1503–1511.
Kobayashi, Y, Horiguchi, R, Miura, S, et al. 2010a. Sex- and tissue- specific expression of p450 aromatase (cyp19a1a) in the yellowtail clownfish *Amphiprion clarkia*. Comp Biochem Physiol, 155A: 237–244.
Kobayashi, Y, Alam, MA, Horiguchi, R, et al. 2010b. Sexually dimorphic expression of gonadotropin sub units in the pituitary of protogynous honeycomb grouper (*Epinephelus merra*): Evidence that follicle-stimulating hormone (FSH) induces gonadal sex change. Biol Reprod, 82: 1030–1036.
Kobayashi, Y, Horiguchi, R, Nozu, R, et al. 2010c. Expression and localization of forkhead transcriptional factor 2 (Foxl2) in the gonads of protogynous wrasse, *Halichoeres trimaculatus*. Biol Sex Difference, 1: 3doi:10.1186/2042-6410-1-3.
Kochakpour, N, Moens, PB. 2008. Sex specific crossover patterns in zebrafish (*Danio rerio*). Heredity, 100: 489–495.
Kojima, Y, Bhandari, RK, Kobayashi, Y, et al. 2008. Sex change of adult initial-phase wrasse *Halichoeres trimaculatus* by estadiol 17 beta treatment. Gen Comp Endocrinol, 156: 628–632.
Kohn, M, Hozel, J, Vogel, W, et al. 2006. Reconstruction of a 450-My-old ancestral vertebrate protokaryotype. Trends Genet, 22: 203–210.
Kokokiris, L, Brusle, S, Kentouri, M, et al. 1999. Sexual maturity and hermaphroditism of the red porgy *Pagrus pagrus* (Teleostei: Sparidae). Mar Biol, 134: 621–629.
Kokokiris, L, Fostier, A, Athanassopoulou, D, et al. 2006. Gonadal changes and blood sex steroids levels during natural sex inversion in the protogynous Mediterranean red porgy *Pagrus pagrus* (Teleostei: Sparidae). Gen Comp Endocrinol, 149: 42–48.

Kolluru, GR, Reznick, DN. 1996. Genetic and social control of male maturation in *Phallichthys quadripunctatus* (Pisces: Poeciliidae). J Exp Biol, 9: 695–715.

Kondo, M, Nanda, I, Hornung, U, et al. 2003. Absence of the candidate male determining gene *dmrt1(Y)* of medaka from other fish species: Cur Biol, 14: 1664–1669.

Kondo, M, Hornung, U, Nanda, I, et al. 2006. Genomic organization of the sex determining and adjacent regions of the sex chromosomes of medaka. Genome Res, 16: 815–826.

Kondo, M, Nanda, I, Schmid, M, et al. 2009. Sex determination and sex chromosome evolution: insights from medaka. Sex Dev, 3: 88–98.

Komatsu, T, Nakumura, S, Nakamura, M. 2006. Masculinization of female golden rabbit fish *Siganus guttatus* using an aromatase inhibitor treatment during sea differentiation. Comp Biochem Physiol, 143c: 402–409.

Koprunner, M, Thisse, C, Thisse, B, et al. 2001. A zebrafish nanos-related gene is essential for the development of primordial germ cells. Genes Dev, 15: 2877–2885.

Koshimizu, E, Imamura, S, Qi, J, et al. 2011. Embryonic senescence and laminopathies in a progeroid zebrafish model. PLoS ONE 6: e17688. doi:10.1371/journal.pone.0017688.

Kossman, H. 1971. Hermaphroditism und Autogamie beim Karpfen. Naturwissenschaften, 6: 328–329.

Koster, RW, Kuhnlein, RP, Wittbrodt, J. 2000. Ectopic sox3 activity elicits sensory placode formation. Mech Dev, 95: 175–187.

Kovacs, B, Egedi, S, Bartfai, R, et al. 2000. Male specific DNA markers from African catfish (*Clarias gariepinus*). Genetica, 110: 267–276.

Kramer, CR, Imbriano, MA. 1997. Neuropeptide Y (NPY) induces gonad reversal in the protogynous bluehead wrasse *Thalassoma bifasciatum* (Teleostei: Labridae). J Exp Zool, 279: 133–144.

Krens, SFG, Corredor-Adamez, M, He, S, et al. 2008. ERK1 and ERK2 MAPK are key regulators of distinct gene sets in zebrafish embryogenesis. BMC Genetics, 9:196 doi:10:1186/1471-2164-9-196.

Kroon, FJ, Munday, PL, Westcott, DA, et al. 2005. Aromatase pathway mediates sex change in each direction. Proc R Soc Lond, 272B: 1399–1405.

Krovel, AV, Olsen, LC. 2002. Expression of a vas: EGFP transgene in primordial germ cells of the zebrafish. Mech Dev, 116: 141–150.

Krovel, AV, Olsen, LC. 2004. Sexual dimorphic expression pattern of a splice variant of zebrafish vasa during gonadal development. Dev Biol, 271: 190–197.

Krumschnabel, G, Podrabsky, JE. 2009. Fish as model systems for the study of vertebrate apoptosis. Apoptosis, 14: 1–21.

Kubista, M. 2008. Emerging real-time PCR applications. Drug Discovery World Summer, 57–66.

Kumar, S, Subramanian, S. 2002. Mutation rates in mammalian genomes. Proc Natl Acad Sci USA, 99: 803–808.

Kurokawa, H, Aoki, Y, Nakamura, S, et al. 2006. Time-lapse analysis reveals different modes of primordial germ cell migration in the medaka *Oryzias latipes*. Dev Growth Diff, 48: 209–221.

Kurokawa, H, Saito, D, Nakamura, S, et al. 2007. Germ cells are essential for sexual dimorphism in the medaka gonad. Proc Natl Acad Sci USA, 104: 16958–16963.

Kusakabe, M, Kobayashi, T, Todo, T, et al. 2002. Molecular cloning and expression during spermatogenesis of a cDNA encoding testicular 11 beta-hydroxylase (P45011beta) in rainbow trout (*Oncorhynchus mykiss*). Mol Reprod Dev, 62: 456–469.

Kusen, JD, Nakazono, A. 1991. Protogynous hermaphroditism in the parrotfish *Calotomus japonicus*. Jap J Ichthyol, 38: 41–45.

Kuwamura, T. 1997. The evolution of parental care and mating systems among Tankanyikan cichlids. pp 59–86. In: H. Kawanabe, M. Hori, M Nagoshi (eds). Parental Care and Mating Systems. Chukyo University Press, Yagota, Nagoya, Japan.

Kuwamura, T, Suzuki, S, Tanaka, N, et al. 2007. Sex change of primary males in a diandric labrid *Halichoeres trimaculatus*: coexistence of protandry and protogyny within a species. J Fish Biol, 10: 1898–1906.

Kwon, JY, Haghpanah, V, Kogson-Haratado, LM, et al. 2000. Masculinization of genetic female Nile tilapia (*Oreochromis niloticus*) by dietary administration of an aromatase inhibitor during sexual differentiation. J Exp Zool, 287: 46–53.

Kwon, JY, Brendan, JM, Penman, DJ, 2001a. Cloning of brain aromatase gene and expression of brain and ovarian aromatase genes during sexual differentiation in genetic male and female Nile tilapia *Oreochromis niloticus*. Mol Reprod Dev, 59: 359–370.

Kwon, J, McAndrew, B, Penman, D. 2001b. Cloning of brain aromatase gene during sexual differentiation in genetic male and female Nile tilapia *Oreochromis niloticus*. Mol Reprod Dev, 62: 456–469.

Lacerda, SMSN, Batlouni, SR, Silva, SBG, et al. 2006. Germ cells transplantation in fish: the Nile tilapia model. Anim Reprod, 3: 146–159.

Lacerda, SMSN, Batlouni, SR, Costa, GMJ, et al. 2010. A new and fast technique to generate offspring after germ cells transplantation in adult fish: The Nile tilapia (*Oreochromis niloticus*) model. PLoS ONE 5:e10740.doi:10.1371/journal. pone.0010740.

Lam, TJ. 1983. Environmental influences on gonadal activity in fish. pp 65–116. In: WS Hoar, DJ Randall, EM Donaldson (eds). Fish Physiology. Academic Press, New York, USA.

Larson, ET, Norris, DO, Grau, EG, et al. 2003. Monoamine stimulate sex reversal in saddleback wrasse. Gen Comp Endocrinol, 130: 289–298.

Lee, MF, Huang, JD, Chang, CF. 2011. Development of the genital duct system in the protandrous black porgy *Acanthopagrus schlegeli*. Anat Rec, 294: 494–505.

Lee, M-F, Huang, J-D, Chang, C-F. 2008. Development of ovarian tissue and female germ cells in the protandrous black porgy *Acanthopagrus schlegeli* (Perciformes: Sparidae). Zool Stud, 47: 302–316.

Lee, Y-H, Du, J-L, Yueh, W-S, et al. 2001. Sex change in the protandrous black porgy *Acanthopagrus schlegeli*: A review in gonadal development, estrodiol, estrogen receptor, aromatase activity and gonadotropin. J Exp Zool, 290: 715–726.

Lee, YM, Seo JS, Kim IC, et al. 2006. Endocrime disrupting chemicals (bisphenol A-4 nonyphenol, 4 tert-octylphenol) modulate expression of hermaphroditic *Rivulus marmoratus*. Biochem Biophys Res Commun, 345: 894–903.

Lengauer, C, Riethman, H, Cremer, T. 1990. Painting of human chromosomes with probes generated from hybrid cell lines by PCR with Alu and L1 primers. Human Genetics, 86: 1–6.

Le Page, Y, Diotel, N, Vaillant, C, et al. 2010. Aromatase, brain sexualization and plasticity: the fish paradigm. Eur J Neurosci, 32: 2105–2115.

Lewis, ZR, McClellan, MC, Postlethwait, JH, et al. 2008. Female-specific increase in primordial germ cells marks sex differentiation in threespine stickleback (*Gasterosteus aculeatus*). J Morphol, 269: 909–921.

Li, GL, Liu, XC, Zhang, Y, et al. 2006. Gonadal development, aromatase activity and p450 aromatase gene expression during sex inversion of protogynous red-spotted grouper *Epinephelus akaara* (Temminck and Schlegel) after initiation of the aromatase inhibitor, Fadrozole. Aquac Res, 37: 484–491.

Li, M, Hong, N, Xu, H, et al., 2009. Medaka *vasa* is required for migration but not survival of primordial germ cells. Mech Dev, 126: 366–381.

Li, Q, Zhou, X, Guo, Y, et al. 2008. Nuclear localization, DNA binding and restricted expression in neural and germ cells of zebrafish Dmrt3. Biol Cell, 100: 453–463.

Li, WH, Yi, S, Makova, K. 2002. Male-driven evolution. Curr Opin Genet Dev, 12: 650–656.

Liarte, S, Chaves-Pozo, E, Garcia-Alcazar, A, et al. 2007. Testicular involution prior to sex change in gilthead seabream is characterized by decrease in DMRT1 gene expression and by massive leucocyte infiltration. Reprod Biol Endocrinol, 5: 20 doi:10.1186/1477-7827-5-20.

Lieske, E, Myers, R. 1996. Coral-Reef Fishes. Princeton University Press, Princeton, New Jersey, USA.
Lindeman, RE, Pelegri, F. 2010. Vertebrate maternal-effect genes: Insights into fertilization, early cleavage divisions and germ cell determinant localization from studies in the zebrafish. Mol Reprod Dev, 77: 299–313.
Linhart, O, Kudo, S. 1997. Surface ultrastructure of paddlefish eggs before and after fertilization. J Fish Biol, 51: 573–582.
Lindholm, AK, Brooks, R, Breden, F. 2004. Extreme polymorphism in a Y-linked sexually selected trait. Heredity, 92: 156–162.
Liu, J-F, Guiguen, Y, Liu, SJ. 2009. Aromatase (p450 arom) and 11β hydroxylase (p450 11β) genes one differentially expressed during the sex change process of the protogynous rice field eel *Monopterus albus*. Fish Physiol Biochem, 35: 511–518.
Liu, M, Sadovy de Mitchesson, Y. 2009. Gonad development during sexual differentiation in hatchery-produced orange-spotted grouper (*Epinephelus coioides*) and humpback grouper (*Cromileptes altivelis*) (Pisces: Serranidae; Epinephelinae). Aquaculture, 287: 191–202.
Liu, SJ, Govoroun, M, D'Cotta, H, et al. 2000. Expression of cytochrome p450 (11 beta) (11 beta-hydroxylase) gene during gonadal sex differentiation and spermatogenesis. J Steroid Biochem Mol Biol, 75: 291–298.
Liu, SJ, Liu, S, Tao, M, et al. 2007. Isolation and expression analysis of testicular type Sox9b in allotetraploid fish. Mar Biotechnol, 9: 329–334.
Liu, Z, Wu, F, Jiao, B, et al. 2007. Molecular cloning of double sex and mab-3-related transcription factor 1, transcription factor gene 2 and two types of cytochrome p450 aromatase in Southern catfish and their possible roles in sex differentiation. J Endoccrinol, 194: 223–241.
Liu, ZH, Zhang, YG, Wang, DS. 2008. Studies on feminization, sex determination and differentiation of the Southern catfish *Silurus meridionalis*. Fish Physiol Biochem, DOI 10.12007/s10695-008-9281-7.
Liu, ZH, Zhang, YG, Wang, DS. 2010. Studies on feminization, sex determination, and differentiation of the Southern catfish, *Silurus meridionalis*—a review. Fish Physiol Biochem 36: 223–235.
Lo Nostro, EL, Guerrero, GA. 1996. Presence of primary and secondary males in a population of the protogynous *Synbranchus marmoratus*. J Fish Biol, 49: 788–800.
Lorenzi, V, Earley, RL, Grober, MS. 2006. Preventing behavioural interactions with a male facilitates sex change in female bluebanded gobies *Lythrypnus dalli*. Behav Ecol Sociobiol, 59: 715–722.
Loukovitis, D, Sarropoulou, E, Tsigenopoulos, CS. 2011. Quantative trait loci involved in sex determination and body growth in the gilthead sea bream (*Sparus aurata* L.) through targeted genome scan. PLoS ONE, 6: e16599-doi:10.1371/journal.pone.0016599.
Lowe, TP, Larkin, JR, 1975. Sex reversal in *Betta splendens* Regan with emphasis on the problem sex determination. J Exp Zool, 191: 25–32.
Luckenbach, JA, Godwin, J, Daniels, HV, et al. 2003. Gonadal differentiation and effects of temperature on sex determination in shallow flounder *Paralichthys lethostigma*. Aquaculture, 216: 315–327.
Luckenbach, JA, Kusakabe, M, Swanson, P, et al. 2008. Unilateral ovariectomy increases egg size and reduces follicular atresia in the semelparous coho salmon *Oncorhynchus kisutch*. J Exp Zool, 309A: 468–476.
Luckenbach, JA, Dickey, JT, Swanson, P. 2010. Regulation of pituitary GnRH receptor and gonadotropin subunits by IGF1 and GnRH in prepubertal male coho salmon. Gen Comp Endocrinol, 167: 387–396.
Luo, Y-S, Hu, W, Liu, X-C, et al. 2010. Molecular cloning and mRNA expression pattern of sox9 during sex reversal in orange spotted grouper (*Epinephelus coioides*). Aquaculture, 306: 322–328.
Ma, H, Chen, S, Yang, J, et al. 2010. Isolation of sex specific AFLP markers in spotted halibut (*Verasper variegatus*). Env Biol Fish, 88: 9–14.

Maack, G, Segner, H. 2003. Morphological development of the gonads in zebrafish. J Fish Biol, 62: 895–906.

Machado, HE, Pollen, AA, Hofmann, HA, et al. 2009. Interspecific profiling of gene expression informed by comparative genomic hybridization. A review and novel approach in African cichlid fishes. Integ Comp Biol, 49: 644–659.

Machado, TC, Pansinato-Alves, JC, Pucci, MB, et al. 2011. Chromosomal painting and ZW sex chromosomes differentiation in *Characidium* (Characiformes; Crenuchidae). BMC Genetics, 12–65 doi:10.118611471-2156-12-65.

Majhi, SK, Hattori, RS, Yokota, M, et al. 2009. Germ cell transplantation using sexually competent fish: An approach for rapid propagation of endangered and valuable germlines. PLoS ONE, 4: e6132.doi:10.1371/journal.pone, 0006132.

Marchand, O, Govoroun, M, D'Cotta, H, et al. 2000. DMRT1 expression during gonadal differentiation and spermatogenesis in the rainbow trout *Oncorhynchus mykiss*. Biochem Biophys Acta, 1493: 180-187.

Martin, E, Taborsky, M. 1997. Alternative male mating tactics in a cichlid, *Pelvicachromis pulcher*: a comparison of reproductive effort and success. Behav Ecol Sociobiol, 41: 311–319.

Martins, RS, Deloffre, LA, Myloans, CC, et al. 2007. Developmental expression of Daxi in the European sea bass *Dicentrarchus labrax*: lack of evidence for sexual dimorphism during sex differentiation. Reprod Biol Endocrinol, 5: 19.

Mathew, G. 2011. Gonadal restructuring during sex transformation in the protogynous greasy grouper *Epinephelus tauvina* (Forsskal) (Peraformes: Serranidae). Indian J Fish, 58: 39–43.

Matsuda, M. 2005. Sex determination in the teleost medaka *Oryzias latipes*. Ann Rev Genet, 39: 293–307.

Matsuda, M, Nagahama, Y, Shinomiya, A, et al. 2002. *Dmy* is a Y specific *Dm*-domain gene required for male development in the medaka fish. Nature, 417: 553–563.

Matsuda, M, Sato,T, Toyazaki, Y, et al. 2003. *Oryzias curvinotus* has *Dmy*, a gene that is required for male development in the medaka, *O. latipes*. Zool Sci, 20: 159–161.

Matsuda, M, Shinomiya, A, Kinoshita, M, et al. 2007. *Dmy* gene induces male development in genetically female (XX) medaka fish. Proc Natl Acad Sci USA, 104: 3865–3870.

Matsuyama, M, Lara, RT, Matsuura, S. 1988. Juvenile bisexuality in the red sea bream *Pagrus major*. Env Biol Fish, 21: 27–36.

Mayer, I, Borg, B, Berglund, I, et al. 1991. Effects of castration and androgen treatment on aromatase activity in the brain of mature male Atlantic salmon (*Salmo salar* L.) parr. Gen Comp Endocrinol, 82: 86–92.

McCafferty, S, Bermingham, E, Quenouille, et al. 2002. Historical biogeography and molecular systematic of the Indo-Pacific genus *Dascyllus* (Teleostei: Pomacentridae). Mol Ecol 11: 1377–1392.

McCormick, CR, Bos, DH, DeWoody, JA. 2008. Multiple molecular approaches yield no evidence for sex-determining genes in lake sturgeon (*Acipenser fulvescens*). J Appl Ichthyol, 24: 643–645.

McCornick, MI, Ryen, CA, Munday, PL, et al. 2010. Different mechanisms underlie sexual size-dimorphism in two populations of a sex changing fish. PLoS ONE, 5: e10616:doi:1371/journal.pone.0010616.

MeiSheng, YI, Hong, Ni, Li, ZD, et al. 2010. Medaka fish stem cells and their applications. Sci China Life Sci, 53: 426–434.

Menuet, A, Pellegrini, E, Anglade, I, et al. 2002. Molecular characterization of three estrogen receptor forms in zebrafish: Binding characteristics, transactivation, properties and tissue distributions. Biol Reprod, 66: 1881–1892.

Micale, V, Perdichizzi, F. 1994. Further studies on the sexuality of the hermaphroditic teleost *Diplodus sargus* with particular reference to protandrous sex inversion. J Fish Biol, 45: 661–670.

Minto, C, Nolan, CP. 2006. Fecundity and maturity of orange roughy (*Hoplostethus atlanticus* Collett, 1889) on the Porcupine Bank, Northeast Atlantic. Env Biol Fish, 77: 39–50.

Miranda, LA, Strussmann, CA, Somoza, GM. 2001. Immunocytochemical identification of GtH1 and GtH2 cells during the temperature-sensitive period of sex determination in pejerrey *Odontesthes bonariensis*. Gen Comp Endocrinol, 124: 45–52.

Miranda, JA, Oliveira, RF, Carneiro, LA, et al. 2003. Neurochemical correlates of male polymorphism and alternative reproductive tactics in Azorean rock-pool blenny *Parablennius parvicornis*. Gen Comp Endocrinol, 132: 181–189.

Miura, T, Miura, C, Konda, Y, et al. 2002. Spermatogenesis preventing substance in Japanese eel. Development, 129: 2689–2697.

Molloy, PP, Goodwin, NB, Cote, IM, et al. 2007. Sperm competition and sex change. A comparative analysis across fishes. Evolution, 61: 640–652.

Mommens, M, Fernamies, JMU, Bizuayehu, T, et al. 2010. Maternal gene expression in Atlantic halibut (*Hippoglossus hippoglossus* L.) and its relation to egg quality. BMC Res Notes, 3: 138.

Monteiro, P, Bentes, L, Coelho, R, et al. 2006. Age and growth, mortality reproduction and relative yield per recruit of the bogue *Boops boops* Linne 1978 (Sparidae) from the Algarve (south of Portugal) longline. J Appl Ichthyol, 22: 345–352.

Morinaga, C, Saito, D, Nakamura, S, et al. 2004. Mutations affecting gonadal development in medaka *Oryzias latipes*. Mech Dev, 121: 829–839.

Morinaga, C, Saito, D, Nakamura, S, et al. 2007. The *hotei* mutation of medaka in the anti-Mullerian hormone receptor causes the dysregulation of germ cell and sexual development. Proc Natl Acad Sci USA, 104: 9691–9696.

Morisawa, S, Cherr, GN, 2002. Acrosome reaction in spermatozoa from hagfish (Agnatha) *Eptatretus burgeri* and *Eptatretus stouti*: Acrosomal exocytosis and identification of filamentous actin. Dev Growth Diff, 44: 337–344.

Morita, K, Takashima, Y. 1998. Effect of female size and fecundity and egg size in white-spotted charr: Comparison between sea-run and resident forms. J Fish Biol, 53: 1140–1142.

Morrish, BC, Sinclair, AH. 2002. Vertebrate sex determination: many means to end. Reproduction, 124: 447–457.

Moser, HG. 2007. Reproduction in the viviparous South African clinid fish *Fuconimus mus*. S Afr J Mar Sci, 29: 423–436.

Mouriec, K, Gueguen, M-M, Manuel, C, et al. 2009. Androgens upregulate *cyp19a1b* (Aromatase B) gene expression in the brain of zebrafish (*Danio rario*) through estrogen receptors. Biol Reprod, 80: 889–896.

Mullis, KB, Faloona, FA. 1987. Specific synthesis of DNA *in vitro* via a polymerase-catalyzed chain reaction. *Methods in Enzymology*, 155: 335–50.

Mullis, KB, Faloona, FA, Scharf, S, et al. 1986. Specific enzymatic amplification of DNA in vitro: the polymerase chain reaction. *ColdSpringHarbor Symposia on Quantitative Biology*, 1986.

Munday, PL, Wilson, W, Warner, RR. 2006. A social basis for the development of primary males in sex changing fish. Proc R Soc Lond, 273B: 2845–2851.

Munehara, H, Takano, K, Koya, Y. 1989. Internal gametic association and external fertilization in the Elkhorn sculpin *Alcichthys alcicornis*. Copeia, 1989: 675–678.

Muriach, B, Cerda-Reverter, JM, Gomez, Z, et al. 2008. Molecular characterization and central distribution of the estrodiol receptor alpha (ER alpha) in the sea bass (*Dicentrarchus labrax*). J Chem Neuroanat, 35: 33–48.

Murray, JD, Maga, EA. 2010. Is there a risk from not using GE animals? Transgenic Res, 19: 357–361.

Murugesan, P, Balasubramanian, T, Pandian, TJ. 2010. Does haemocoelom exclude embryonic stem cells and asexual reproduction in invertebrates? Curr Sci, 98: 768–771.

Nagai, T, Takehana, Y, Hamaguchi, S, et al. 2008. Identification of the sex determining locus in the Thai medaka *Oryzias minutillus*. Cytogenet Genome Res, 121: 137–142.

Nagler, JJ, Cavileer, T, Steinhorst, K, et al. 2004. Determination of sex in chinook salmon (*Oncorhynchus tshawytscha*) using male-linked growth hormone pseudogene by Real-Time PCR. Mar Biotechnol, 6: 186–191.

Nagler, JJ, Cavileer, T, Sullivan, D, et al. 2007. The complete nuclear estrogen receptor family in the rainbow trout: discovery of the novel ER alpha2 ad both ER beta isoforms. Gene, 392: 164–173.

Nagler, JJ, Cavileer, T, Hunter, S, et al. 2011. Non-sex specific genes associated with the secondary mitotic period of primordial germ cell proliferation in the gonads of embryonic rainbow trout (*Oncorhynchus mykiss*). Mol Reprod Dev, Doi 10.1002/mrd.21277.

Nakamoto, M, Kobayashi, T, Chang, X, et al. 1998. Gonadal sex differentiation in teleost fish. J Exp Zool, 281: 362–372.

Nakamoto, M, Suzuki, A, Matsuda, M, et al. 2005. Testicular type of Sox9 is not involved in sex determination but might be in the development of testicular structures in the medaka *Oryzias latipes*. Biochem Biophys Res Commun, 333: 729–736.

Nakamoto, M, Matsuda, M, Wang, D, et al. 2006. Molecular cloning and analysis of gonadal expression of Foxl2 in the medaka *Oryzias latipes*. Biochem Biophys Res Commun, 344: 353–361.

Nakamura, M. 1984. Effects of estrodiol 17-beta on gonadal sex differentiation in two species of salmonids, the masou salmon *Oncorhynchus masou* and the chum salmon *O. keta*. Aquaculture 43: 1–3.

Nakamura, M, Kobayashi, T, Chang, XT, et al. 1998. Gonadal sex differentiation in teleost fish. J Exp Zool, 281: 362–372.

Nakamura, M, Bhandari, RK, Higa, M. 2003. The role estrogens play in sex differentiation and sex changes of fish. Fish Physiol Biochem, 28: 113–117.

Nakamura, S, Kobayashi, D, Aoki, Y, et al. 2006. Identifications and lineage of populations of somatic gonadal precursors in medaka embryos. Dev Biol, 295: 678–688.

Nakamura, S, Aoki, Y, Saito, D, et al. 2008. Sox9b/sox9a2-EGFP transgenic medaka reveals the morphological reorganization of the gonad and a common precursor of both female and male supporting cells. Mol Reprod Dev, 75: 472–476.

Nakamura, S, Kobayashi, K, Nishimura, T, et al. 2010. Identification of germline stem cells in the ovary of the teleost medaka. Science, 328: 1561–1563.

Nakamura, S, Kobayashi, K, Nishimura, T, et al. 2011. Ovarian germline stem cells in the teleost fish medaka (*Oryzias latipes*). Int J Biol Sci, 7: 403–409.

Nakamura, Y, Suga, K, Sakakura, Y, et al. 2008. Genetic and growth differences in the outcrossings between two clonal strains of the self-fertilizing mangrove killifish. Can J Zool, 86: 976–982.

Nakayama, I, Biagi, CA, Koide, N, et al. 1998. Identification of a sex-linked *GH* pseudogene in two species of Japanese salmon (*Oncorhynchus masou* and *O. rhodurus*). Aquaculture, 173: 65–72.

Nakazono, A, Kuwamura, T. 1987. Sex change in fishes. Tokai University Press, Tokyo, Japan.

Nakazono, A, Kusen, JD. 1991. Protogynous hermaphroditism in the wrasse *Choerodon azurio*. Nippon Suisan Gakkaishi, 57: 417–420.

Nanda, I, Schartl, M, Feichtinger, W, et al. 1992. Early stages of sex chromosome differentiation in fish as analysed by simple repetitive DNA sequence. Chromosoma, 101: 301–310.

Nanda, I, Volff, JN, Weis, S, et al. 2000. Amplification of a long terminal repeat like element on the Y chromosome of the platyfish *Xiphophorus maculatus*. Chromosoma, 109: 173–180.

Nanda, I, Kondo, M, Hornung, U, et al. 2002. A duplicated copy of *DMRT 1* in the sex-determining region of the Y chromosome of the medaka *Oryzias latipes*. Proc Natl Acad Sci USA, 99: 11778–11783.

Nanda, I, Hornung, U, Kondo, M, et al. 2003. Common spontaneous sex-reversed XX males of the medaka *Oryzias latipes*. Genetics, 163: 245–251.

Neff, BD. 2003a. Genetic paternity analysis and breeding success in bluegill sunfish (*Lepomis macrochirus*). J Hered, 92: 111–119.

Neff, BD. 2003b. Paternity and condition affect cannibalistic behavior in nest-bending bluegill sunfish. Behav Ecol, 32: 229–237.

Nunez, J, Duponchelle, F. 2009. Towards a universal scale to assess sexual maturation and related life history in oviparous teleost fishes. Fish Physiol Biochem, 35: 167–180.

Ogino, Y, Katoh, H, Yamada, G. 2004. Androgen dependent development of a modified anal fin, genopodium as a model to understand the mechanism of secondary sexual character expression in vertebrates. FEBS Lett, 575: 119–126.

Ogino, Y, Miyagawa, S, Katoh, H, et al. 2011. Essential functions of androgen signalling emerged through the developmental analysis of vertebrate sex characteristics. Evol Dev, 13: 315–325.

Ohta, K, Sundaray, JK, Okida, T, et al. 2003. Bidirectional sex change and its steroidogenesis is in the wrasse *Pseudolabrus sieboldi.* Fish Physiol Biochem, 28: 173–174.

Okada, YK, Yamashita, H. 1994. Experimental investigation of the manifestation of secondary sexual characters in the fish using medaka *Oryzias latipes* as material. J Fac Sci Tokyo Univ, Tokyo Sect IV, 6: 383–347.

Okutsu, T, Suzuki, K, Takeuchi, Y, et al. 2006. Testicular germ cells can colonize sexually undifferentiated embryonic gonad and produce functional egg in fish. Proc Natl Acad Sci USA, 103: 2725–2729.

Okutsu, T, Shikana, S, Kanno, M, et al. 2007. Production of trout offspring from triploid salmon parents. Science, 317: 1517.

Okutsu, T, Takeuchi, Y, Yoshizaki, G, et al. 2008. Spermatogonial transplantation in fish: Production of trout offspring from salmon parents pp. 209–219. In: K Tsukamoto, T Kawamura, T Takeuchi, TD Beard, Jr, MJ Kaiser (eds) Fish Global Welfare Environ, 5th World Fish Cong, 2008.

Oliveira, RF. 2006. Neuroendocrime mechanisms of alternative reproduction tactics in fish. pp. 297–357. In: WS Hoar, DJ Randall (eds) Fish Physiology, vol. 24 Behaviour and Physiology. Academy Press, New York, USA.

Oliveira, RF, Canario, AVM, Grober, MS, et al. 2001a. Endocrime correlates of male polymorphism and alternative reproductive tactics in the Azorean rock-pool blenny *Parablennius sanguinolentus parvicornis*. Gen Comp Endocrinol, 121: 278–288.

Oliveira, RF, Carneiro, LA, Canario, AVM, et al. 2001b. Effects of androgens on social behavior and morphology of alternative reproductive males of the Azorean rock-poll blenny. Horm Behav, 39: 157–166.

Oliveira, RF, Carneiro, LA, Gonclaves, DM, et al. 2001c. 11-ketotestosterone inhibits the alternative mating tactics in sneaker males of the peacock blenny *Salaria pavo*. Brain Behav Evol, 158: 28–37.

Oliveira, C, Foresti, F, Hilsdorf, HWS. 2009. Genetics of Neotropical fish: from chromosomes to populations. Fish Physiol Biochem, 35: 81–100.

Olsen, E, Lilly, GR, Heino, M, et al. 2005. Assessing changes in age and size at maturation in collapsing populations of Atlantic cod (*Gadus morhua*). Can J Fish Aquat Sci, 62: 811–823.

Olsen, LC, Aasland, R, Fjose, AA. 1997. *vasa*-like gene in zebrafish identifies putative primordial germ cells. Mech Dev, 66: 95–105.

Olsson, PE, Berg, AH, von Hofsten, B, et al. 2005. Molecular cloning and characterization of a nuclear androgen receptor activated by 11-ketotestosterone. Reprod Biol Endocrinol, 3: Article No 37.

Onitake, K. 1972. Morphological studies of normal sex differentiation and induced sex reversal process of gonads in the medaka *Oryzias latipes*. Anno Zool Japan, 45: 159–169.

Orlando, EF, Katsu, Y, Miyagawa, S, et al. 2006. Cloning and differential expression of estrogen receptor and aromatase genes in the self fertilizing hermaphrodite and male mangrove rivulus *Kryptolebias marmoratus*. J Mol Endocrinol, 37: 353–365.

Otake, H, Shinomiya, A, Matsuda, M, et al. 2006. Wild-derived XY sex reversal mutants in the medaka, *Oryzias latipes*. Genetics, 173: 2083–2090.

Otake, H, Hayashi, Y, Hamaguchi, S, et al. 2008. The Y chromosome that lost the male-determining function behaves as an X chromosome in the medaka fish *Oryzias latipes*. Genetics, 179: 2157–2162.

Oteiza, P, Koppen, M, Concha, ML, et al. 2008. Origin and shaping of the laterality organ in zebrafish. Development, 135: 2807–2813.
Pajuelo, JG, Lorenzo, JM, Bilbao, A, et al. 2006. Reproductive characteristics of the benthic coastal fish *Diplodus vulgaris* (Teleostei: Sparidae) in the Canarian archipelago, northwest Africa. J Appl Ichthyol, 22: 414–418.
Pala, I, Schartl, M, Thorsteinsdottir, S, et al. 2009. Sex determination in the *Squalius alburnoides* complex: An initial characterization of sex cascade element in the contexts of a hybrid polyploidy genome. PLoS ONE, 4e6401.doi:10.1371/journal pone. 0006401.
Palmitter, RD, Brinster, RL, Hammer, RE, et al. 1982. Dramatic growth of mice that develop from eggs microinjected with metallothionein-growth hormone fusion genes. Nature, 300: 611–615.
Pandian, TJ. 2003. Transplanting the fish. Curr Sci, 85: 101.
Pandian, TJ. 2010. Sexuality in Fishes. Science Publishers Enfield, CRC Press Boca Raton, USA, p. 208.
Pandian, TJ. 2011. Sex Determination in Fish. Science Publishers Enfield/CRC Press Boca Raton, USA, p. 282.
Pandian, TJ. 2012. Endocrine Sex Differentiation in Fish. Science Publisher, Enfield/CRC Press, Roca Baton, USA.
Pandian, TJ, Marian, LA. 1994. Problems and prospects of transgenic fish production. Curr Sci, 66: 635–648.
Pandian, TJ, Sheela, SG. 1995. Hormonal induction of sex reversal in fish. Aquaculture, 138: 1–22.
Pandian, TJ, Koteeswaran, R. 1998. Ploidy induction and sex control in fish. Hydrobiologia, 384: 167–243.
Pandian, TJ, Koteeswaran, R. 1999. Lability of sex differentiation in fish. Curr Sci, 76: 580–583.
Pandian, TJ, Kirankumar, S. 2003. Androgenesis and conservation of fishes. Curr Sci, 85: 917–931.
Pandian, TJ, Koteswaran, R, Vivekanandan, E. 2012. Environmental Sex Determination and Differentiation in Fish. Science Publisher, Enfield/CRC Press, Roca Baton, USA.
Parameswaran, V, Shukla, R, Bhonde, R, et al. 2007. Development of a pluripotent 'ES-like' cell line from Asian sea bass (*Lates calcarifer*)—an oviparous stem cell line mimicking viviparous ES cells. Mar Biotechnol, 9: 766–775.
Pardue, ML, Gall, LG. 1969. Molecular hybridization of radioactive DNA to the DNA of cytological preparations. Proc Natl Acad Sci USA, 64: 600–604.
Patnaik, BK, Mahapatro, N, Jena, BS. 1994. Ageing in fishes. Gerontology, 40: 113–132.
Paul-Prasanth, B, Nakamura, M, Nagahama, Y. 2011. Sex Determination in Fishes. pp 1–14. In: DO Norris, KH Lopez (eds) Hormones and Reproduction of Vertebrates, vol 1. Elsevier, Amsterdam, The Netherlands.
Pears, RJ, Choat, JH, Mapstone, BD, et al. 2007. Reproductive biology of a large, aggregation-spawning serranid *Epinephelus fuscoguttatus* (Forskal): Management implications. J Fish Biol, 71: 795–817.
Peichel, CL, Ross, JA, Matson, M, et al. 2004. The master sex-determination locus in three spine sticklebacks is on a nascent Y chromosome. Curr Biol, 14: 1416–1424.
Perner, B, Englert, C, Bolhg, F. 2007. The Wilms tumor genes Wt1a and Wtb control different steps during formation of the zebrafish pronephros. Dev Biol, 309: 87–96.
Peterson, CW. 1991. Variation in fertilization rate in the tropical reef fish *Halichoeres bivittatus*: correlates and implications. Biol Bull, 181: 232–237.
Piferrer, F, Blazquez, M. 2005. Aromatase distribution and regulation in fish. Fish Physiol Biochem, 31: 215–226.
Piferrer, FC, Guiguen, Y. 2008. Fish gonadogenesis, part II. Molecular biology and genomics of sex differentiation. Rev Fish Sci, 16(S1): 35–55.
Piferrer, F, Carillo, M, Zanuy, S, et al. 1994. Induction of sterility in coho salmon (*Oncorhynchus kisutch*) by androgen immersion before first feeding. Aquaculture, 119: 403–423.

Podrabsky, JE, Lozez, JP , Fan TWM, et al. 2007. Extreme anoxia tolerance in embryos of the annual killifish *Austrofundulus linnaeus*: insights from a metabolic analysis. J Exp Biol, 210: 2253–2266.

Psenicka, M, Alavi, SMH, Rodina, M, et al. 2007. Morphology and ultrastructure of Siberian sturgeon *Acipenser beari* spermatozoa using scanning and transmission electron microscopy. Biol Cell, 99: 103–115.

Psenicka, M, Alavi, SMH, Vancova, M, et al. 2009. Fine structure and morphology of sterlet (*Acipenser ruthenus* L. 1958) spermatozoa and acrosin localization. Anim Reprod Sci, 111: 3–16.

Psenicka, M, Rodina, M, Linhart, O. 2010. Ultrastructure study on the fertilization process in sturgeon (*Acipenser*) function of acrosome and prevention of polyspermy. Anim Reprod Sci, 117: 147–154.

Psenicka, M, Kaspar, V, Alavi, SMH, et al. 2011. Potential role of the acrosome sturgeon spermatozoa in the fertilization process. J Appl Ichthyol, 27: 678–682.

Raghuveer, K, Senthilkumaran, B. 2009. Identification of multiple dmrt1 in catfish: localization, dimorphic expression pattern, changes during testicular cycle and after methyltestosterone treatment. J Mol Endocrinol, 42: 437–448.

Raghuveer, K, Senthilkumaran, B, Sudhakumari, CC, et al. 2011. Dimorphic expression of various transcription factor and steroidogenic enzyme genes during gonadal ontogeny in the air-breathing catfish *Clarias gariepinus*. Sex Dev, 5: 213–223.

Rattaunayuvakorn, S, Mungkornkarn, P, Thongpan, A, et al. 2006. Gonadal development and sex immersion in saddleback anemonefish *Amphiprion polymous* Linnaeus (1758). Kasetsart J (Nat Sci), 40: 196–203.

Ravaglia, MA, Maggese, MC. 2002. Oogenesis in the swamp eel *Synbranchus marmoratus* (Bloch 1975) (Teleostei: Synbranchidae). Ovarian anatomy stages of oocyte development and micropyle structure. Biocell, 26: 325–337.

Reinboth, R. 1972. Some remarks on secondary sex characters, sex and sexual behavior in teleosts. Gen Comp Endocrinol, S 3: 565–570.

Reinboth, R, Brusle-Sigard, S. 1997. Histological and ultrastructural studies on the effects of hcG on sex inversion in the protogynous teleost *Coris julis*. J Fish Physiol, 51: 733–749.

Reznick, D, Bryant, M, Holmes, D. 2006. Evolution of senescence and post-reproductive lifespan in guppies (*Poecilia reticulata*). PLoS Biol, 4: DOI: 10.1371/journal.pbio.0010007.

Rhee, JS, Seo, JS, Raisuddin, S, et al. 2008. Gonadotropin releasing hormone receptor (GnRHR) gene expression is differently modulated in gender types of the hermaphroditic fish *Kryptolebias marmoratus* by endocrine disrupting chemicals. Comp Biochem Physiol Toxicol Pharmocol, 147: 357–365.

Rhee, JS, Kim, BM, Lee, CJ, et al. 2011. Bisphenol A modulates expression of sex differentiation genes in the self fertilizing fish *Kryptolebias marmoratus*. Aquat Toxicol, 104: 218–219.

Riehl, R, Spatzner, RA, 1991. Breeding, egg structure and larval morphology of the catfish *Sturisoma aureum* (Steindachner) (Teleostei, horecariidae). J Aquacult Aquat Sci, 6: 1–6.

Riehl, R, Kokoscha, M. 1993. A unique surface pattern and micropylar apparatus in the eggs of *Luciocephalus* sp (Perciformes, Loricocephalidae). J Fish Biol, 43: 617–620.

Robertson, OH. 1961. Prolongation of the life span of kokanee salmon (*Oncorhynchus nerka kennerlyi*) by castration before beginning of gonad development. Proc Natl Acad Sci USA, 47: 609–611.

Rodriquez-Mari, A, Yan, YL, BreMiller, RA, et al. 2005. Characterization and expression pattern of zebrafish Anti-Mullerian hormone (*amh*) relative to sox9a, sox9b and *cyp19a1a* during gonadal development. Gene Expr Patterns, 5: 655–667.

Rodriquez-Mari, A, Canestro, C, BreMiller, RA, et al. 2010. Sex reversal in zebrafish *fancl* mutants is caused by Tp 53-mediated germ cell apoptosis. PLoS Genet, 6: e1001034:doi:10.137/journal.pgen. 1001034.

Roff, DA. 1992. The Evolution of Life Histories. Chapman and Hall, London, UK.

Ross, RM. 1990. The evolution of sex change mechanisms in fishes. Env Biol Fish, 29: 81–93.

Ross, JA, Peichel, CL. 2008. Molecular cytogenetic evidence of rearrangements on the Y chromosome of the threespine stickleback fish. Genetics, 179: 2173–2182.

Ross, JA, Urton, JR, Boland, J, et al. 2009. Turnover of sex chromosomes in the stickleback fishes (Gasterosteidae). PLoS Genet, 5:e1000391.

Routray, P, Dash, C, Dash, SN, et al. 2010. Cryopreservation of isolated blastomeres and embryonic stem-like cells of leopard danio, *Brachydanio frankei.* Aquacult Res, 41: 579–589.

Ruis-Carus, R. 2002. Chromosome analysis of the sexual phases of the protogynous hermaphrodites *Epinephelus guttatus* and *Thalassoma bifasciatum.* (Serranidae, Labridae, Teleostei). Carrib J Sci, 38: 44–51.

Ryan, MJ, Causey, BA. 1989. "Alternative" mating behavior in the swordtails *Xiphophorus nigrensis* and *Xiphophorus pygmaeus* (Pisces: Poeciliidae). Behav Ecol Sociobiol, 24: 341–348.

Ryu, S, Holzschuh, J, Erhardt, S, et al. 2005. Depletion of minichromosome maintenance protein 5 in the zebrafish retina cause cell cycle defect and apoptosis. Proc Natl Acad USA, 102: 18467–18472.

Sabbah, S, Laria, RL, Gray, SM, et al. 2010. Functional diversity in the color vision of cichlid fishes. BMC Biol, 8: 133–148.

Sadler, TW. 2004. Langman's Medical Embryology. Lippincot Williams & Wilkins, Baltimore, USA.

Sadovy, Y, Shapiro, DY. 1987. Criteria for the diagnosis of hermaphroditism in fishes. Copeia, 1987: 136–156.

Sadovy, Y, Colin, PL. 1995. Sexual development and sexuality in the Nassau grouper. J Fish Biol, 46: 961–976.

Sadovy de Mitcheson, Y, Liu, M. 2008. Functional hermaphroditism in teleosts. Fish Fish, 9: 1–43.

Saiki, R, Gelfand, D, Stoffel, S, et al. 1988. Primer-directed enzymatic amplification of DNA with a thermostable DNA polymerase. Science 239(4839): 487–491.

Saito, D, Tanaka, M. 2009. Comparative aspects of gonadal sex differentiation in medaka: A conserved role of developing oocytes in sexual canalization. Sex Dev, 3: 99–107.

Saito, D, Morinaga, C, Aoki, Y, et al. 2007. Proliferation of germ cells during gonadal sex differentiation in medaka. Insights from germ cell-depleted mutant *zenzai.* Dev Biol, 310: 280–290.

Saito, T, Goto-Kazeto, R, Arai, K, et al. 2008. Xenogenesis in teleost fish through generation of germ line chimeras by single primordial germ cell transplantation. Biol Reprod, 78: 159–166.

Saito, T, Goto-Kazeto, R, Fujimoto, T, et al. 2010. Inter-species transplantation and migration of primordial germ cells in cyprinid fish. Int J Dev Biol, 54: 1481–1486.

Sakakura, Y, Noakes, DLG. 2000. Age, growth, sexual development in the self fertilizing hermaphroditic fish *Rivulus marmoratus.* Env Biol Fish, 59: 309–317.

Sambrook, J, Russell, DW. 2001. Rapid amplification of 5′ cDNA ends. In: Molecular Cloning: A Laboratory Manual Cold Spring Harbor Laboratory Press, Cold Spring Harbor, New York, USA 8: (9), 54–60.

Sandra, G-E, Norma, M-M. 2009. Sexual determination and differentiation in teleost fish. Rev Fish Biol. Fisher, Doi 10.1007/s11160.009-9123-4.

Sanetra, M, Henning, F, Fukaniachi, S, et al. 2009. A microsatellite-based genetic linkage map of the cichlid fish, *Astatotilapia burtoni* (Teleoste). A comparision of genomic architectures among rapidly speciating cichlids. Genetics, 182: 387–397.

Santos, EM, Workmann, VL, Paull, GC, et al. 2007. Molecular basis of sex and reproductive status in breeding zebrafish. Physiol Genomics, 30: 111–122.

Sawatari, E, Shikina, S, Takeuchi, T, et al. 2007. A novel transforming growth factor-beta super family member expressed in gonadal somatic cells enhances primordial germ cell and spermatogonial proliferation in rainbow trout (*Oncohynchus mykiss*). Dev Biol, 301: 266–275.

Sawyer, SJ, Gerstner, KA, Callard, GV. 2006. Real-time PCR analysis of cytochrome p450 aromatase expression in zebrafish: gene specific tissue specific distribution sex differentiation, development programming and estrogen regulation. Gen Comp Endocrinol, 147: 108–117.

Schaefer, SA. 1998. Conflict and resolution iompact of new taxa in phylogenetic studies of the Neotropical cascudinhos (Siluroidei, Labricariidae). In: LR Malabarba, RE Reis, PP Vari (eds) Phylogeny and Classification of Neotropical Fishes. Edupucrs, Porto Alegre, Brazil.

Scholz, S, Rosler, S, Schaffer, M, et al. 2003. Hormonal induction and stability of monosex populations in the medaka (*Oryzias latipes*): Expression of sex specific marker genes. Biol Reprod, 69: 673–678.

Schreibman, MP, Kallman, KD. 1977. The genetic control of the pituitary-gonadal axis in the platyfish, *Xiphophorus maculatus*. J Exp Zool, 200: 277–294.

Schreibman, MP, Kallman, KD, 1978. The genetic control of sexual maturation in the teleost *Xiphophorus maculatus* (Poeciliidae): a review. Ann Biol Bioch Biophy 18: 957–962.

Schultheis, C, Zhou, Q, Froschauer, A, et al. 2006. Molecular analysis of the sex-determining region of the platyfish *Xiphophorus maculatus*. Zebrafish, 3: 299–309.

Schultheis, C, Bohne, A, Schartl, M, et al. 2009. Sex determination diversity and sex chromosome evolution in poeciliid fish. Sex Dev, 3: 68–77.

Schulz, H. 1996. Drastic decline of the proportion of males in the roach (*Rutilus rutilus* L.) population of Bautzen Reservoir (Saxony, Germany): Result of direct and indirect effect of bioaccumulation. Limnologia, 26: 153–164.

Schwarz, AL, Smith, CA. 1990. Sex change in the damselfish *Dascyllus reticulatus* (Richardson) (Pereiformes: Pomacentridae). Biol Mar Sci, 46: 790–798.

Ser, JR, Roberts, RB, Kocher, TD. 2009. Multiple interacting loci control sex determination in Lake Malawi cichlid fish. Evolution doi: 10.1111/j.1558.5646.2009: 00871.x.

Shapiro, DY, Rasotta, MB. 1993. Sex differentiation and gonadal development in the diandric protogynous wrasse *Thalassoma bifasciatum* (Pisces, Labridae). J Zool Lond, 230: 231–245.

Shapiro, DY, Marconato, A, Yoshikawa, T. 1994. Sperm economy of a coral reef fish. Ecology, 75: 1334–1344.

Shapiro, MD, Summers, BR, Balabhadrov, S, et al. 2009. The genetic architecture of skeletal convergence and sex determination in ninespine sticklebacks. Curr Biol, 19: 1140–1145.

Shen, X-Y, Cui, J-Z, Gong, Q-L, et al. 2007. Cloning of the full-length coding sequence and expression analysis of Sox9b in guppy (*Poecilia reticulata*). Fish Physiol Biochem, 33: 195–202.

Shi, B, Wen, HS, Shuang, FH, et al. 2009. Association of reproductive performance with SNPs of FoxL2 gene by SSCP in Japanese flounder (*Paralichthys olivaceus*). Comp Biochem Physiol, 153B: 1–7.

Shibuno, T, Chiba, I, Hashimoto, et al. 1994. Reproductive behavior of the wrasse *Thalassoma lutescens* at Kuchierabu-jima. J Fac Appl Biol Sci, 33: 43–50.

Shimada, A, Takeda, H. 2008. Production of a maternal zygotic medaka using hybrid sterility. Dev Growth Diff, 50: 421–426.

Shinomiya, A, Matsuda, M, Hamaguchi, S, et al. 1999. Identification of genetic sex of the medaka *Oryzias latipes*. Medaka, 10: 31–32.

Shinomiya, AI, Shibata, N, Sakaizumi, M, et al. 2002. Sex reversal of genetic females (XX) induced by the transplantation of XY somatic cells in the medaka *Oryzias latipes*. Int J Dev Biol, 46: 711–717.

Shinomiya, A, Otake, H, Togashi, K, et al. 2004. Field survey of sex reversals in the medaka *Oryzias latipes*: genotypic sexing of wild populations. Zool Sci, 21: 613–619.

Shiraishi, E, Yoshinaga, N, Miura, T, et al. 2008. Mullerian inhibiting substance is required for germ cell proliferation during early gonadal differentiation in medaka (*Oryzias latipes*). Endocrinology, 149: 1813–1819.

Shirak, A, Serrousi, E, Gnaani, A, et al. 2006. Amh and Dmrt 2 genes map to tilapia (*Oreochromis sp*) linkage group 23 within quantitative trait locus regions for sex determination. Genetics, 174: 1573–1581.

Shive, RH, West, RR, Embree, LJ, et al. 2010. *brca2* in zebrafish ovarian development, spermatogenesis and tumorigenesis. Proc Natl Acad Sci USA, 107: 19350–19355.

Shpigel, M, Fishelson, L. 1986. Behavior and physiology of coexistence in two species of *Dascyllus* (Pomacentridae, Teleostei). Env Biol Fish, 17: 253–265.

Siegfried, KR, 2010. In search of determinants: genes expression during gonadal sex differentiation. J Fish Biol, 76: 1879–1902.

Siegfried, KR, Nusslein-Volhard, C. 2008. Germ cell line control of female sex determination in zebrafish. Dev Biol, 324: 277–287.

Silva, MT, doVale, A, dosSantos, NM. 2008. Fish and apoptosis: studies in disease and pharmaceutical design. Curr Pharm Design, 14: 170–183.

Simpson, ER, Mahendroo, MS, Means, GD. 1994. Aromatase cytochrome p450, the enzyme responsible for estrogen synthesis. Endcrinol Rev, 15: 342–355.

Skaar, KS, Norbrega, RH, Magaraki, A, et al. 2010. Proteolytically activated recombinant anti-Mullerian hormone inhibits androgen secretion, proliferation and differentiation of spermatogonia in adult zebrafish testis organ culture. Endocrinology, 152: doi:10.1210/en.2010.1469.

Slanchev, K, Stebler, J, Cueva-Mendez, de la, G, et al. 2005. Development without germ cells: The role of the germ line in zebrafish sex differentiation. Proc Natl Acad Sci USA, 102: 4074–4079.

Small, CM, Carney, GE, Mo, Q, et al. 2009. A microarray analysis of sex- and gonad-biased gene expression in the zebrafish: Evidence for masculinization of the transcriptome. BMC Genomics, 10: 579–594.

So, WK, Kwok, HF, Ge, W. 2005. Zebrafish gonadotropin and their receptors. II Cloning and characterization of zebrafish follicle-stimulating hormone and lutinizing hormone sub-units -their spatial-temporal expression patterns and receptor specificity. Biol Reprod, 76: 1382–1396.

Socorroa, S, Martinsb, RS, Deloffreb, L, et al. 2007. A cDNA for European sea bass (*Dicentrarchus labrax*) 11β-hydroxylase: gene expression during the thermosensitive period and gonadogenesis. Gen Comp Endocrino, 150: 164–173.

Sola, I, Marzovillo, M, Rossi, A, et al. 1997. Cytogenetic analysis of self-fertilizing fish *Rivulus marmoratus*: remarkable chromosomal constancy over a vast geographic range. Genome, 40: 945–949.

Solnica-Krezel, L, Schier, AF, Driever, W. 1994. Efficient recovery of ENU-induced mutations from the zebrafish germline. Genetics, 136: 1401–1420.

Stearns, SC, Crandall, RE. 1984. Plasticity of age and size at sexual maturity: a life history response to unavoidable stress. pp. 13–33. In: GW Potts, RJ Wootton (eds) Fish Reproduction: Strategies and Tactics. Academic Press, London, UK.

Stein, J, Reed, KM, Wilson, CC, et al. 2002. A sex linked microsatellite locus isolated from the Y chromosome of the lake charr *Salvelinus namaycusch*. Env Biol Fish, 64: 211–216.

Strasser, M J, MacKenzie, N C, Dumstrei, K, et al. 2008. Control over the morphology and segregation of zebrafish germ cell granules during embryonic development. BMC Dev Biol, 8: 58.

Strussmann, CA, Nakamura, M. 2003. Morphology, endocrinology and environmental modulation of gonadal sex differentiation in teleost fishes. Fish Physiol Biochem, 26: 13–29.

Strussmann, CA, Moriyama, S, Hanke, EF, et al. 1996. Evidence of thermolabile sex determination in pejerray. J Fish Biol, 48: 643–651.

Strussmann, CA, Saito, T, Takashima, F. 1998. Heat-induced germ cell deficiency in the teleost *Odontesthes bonariensis* and *Patagonian hatcheri*. Comp Biochem Physiol, 119: 637–644.

Strussmann, CA, Kitahara, A, Yamashita, M. 2008. Role of apoptosis in temperature dependent sex determination of pejerray *Odontesthies bonariensis*. Cybium, 32: 77–79.

Sudback, P, Schmits, ML, Baewerle, PA, et al. 1996. Sex reversal by loss of the C-terminal transactivation domain human *Sox9*. Nat Genet, 13: 230–232.

Sudhakumari, CC, Kobayashi, T, Kjiura-Kobayashi, H, et al. 2005. Ontogenic expression patterns of several nuclear receptors and cytochrome p450 aromatases in brain and gonads of Nile tilapia *Oreochromis niloticus* suggests their involvement in sex differentiation. Fish Physiol Biochem, 31: 129–135.

Suzuki, N. 1997. Histological study on intersexual gonads observed in wild tiger puffer *Takifugu rubripes*. Bull Nansei Natl Fish Res Int, 30: 101–113.

Suzuki, A, Shibata, N. 2004. Developmental process of genital ducts in the medaka *Oryzias latipes*. Zool Sci, 21: 397–406.

Suzuki, A, Saga, Y. 2008. *Nanos 2* suppresses meiosis and promotes male germ cell differentiation. Genes Dev, 22: 430–435.

Suzuki, A, Tanaka, M, Shibata, N. 2004. Expression of *aromatase* mRNA and effects of aromatase inhibitor during developmenta in medaka. J Exp Zool, 301A: 266–273.

Suzuki, A, Nakamoto, M, Kato, Y, et al. 2005. Effect of estradiol 17-β on germ cell proliferation and *Dmy* expression during early differentiation in medaka *Oryzias latipes*. Zool Sci, 22: 791–996.

Suzuki, H, Maegawa, S, Nishibu, T, et al. 2000. Vegetal localization of the maternal RNA encoding an EDEN-BP/Bruno like protein in zebrafish. Mech Dev, 93: 205–209.

Svardson, G, Wiokbom, G. 1942. The chromosomes of two species of Anabantidae (Teleostei) with a new case of sex reversal. Heriditas, 28: 212–216.

Taborsky, M. 2001. The evolution of bourgeois, parasitic and cooperative reproductive behaviours of fishes. J Hered, 92: 100–110.

Tagami, T, Kagami, H, Matsubara, Y, et al. 2007. Differentiation of female primordial germ cells in the male testes of chicken (*Gallus gallus domesticus*). Mol Reprod Dev, 74: 68–78.

Takahashi, H. 1977. Juvenile hermaphroditism in the zebrafish *Brachydanio rerio*. Bull Fac Fish Hokkaido Univ, 28: 57–63.

Takahashi, H, Shimizu, M. 1983. Juvenile intersexuality in a cyprinid fish, the Sumutra barb *Barbus tetrazona tetrazona*. Bull Fac Fish Hokkaido Univ, 34: 69–78.

Takahashi, H, Maeno, Y. 1986. Occurrence of an accessory cavity in the testis of the topmouth gudgeon *Pseudorasbora parva* and its relation to juvenile hermaphroditism. Jap J Ichthyol, 33: 180–185.

Takahashi, H, Kobayashi, K, Suzuki, K. 1989. Gonadal morphology of threadfin breams *Nemipterus bathybius* and *N. virgatus:* Evidence of rudimentary hermaphroditism. Jap J Ichthyol, 36: 82–89.

Takamoto, G, Seki, S, Nakashima, Y, et al. 2003. Protogynous sex change in the haremic triggerfish *Sufflamen chrysopterus* (Tetradontoformes). Ichthyol Res, 50: 281–283.

Takehana, Y, Naruse, K, Hamaguchi, S, et al. 2007. Evolution of ZZ/ZW and XX/XY sex determination systems in the closely related medaka species (*Oryzias hubbsi* and *O. dancena*). Chromosoma, 116: 463–470.

Takeo, J, Yamashita, S. 2000. Rainbow trout androgen receptor-alpha fails to distinguish between any of the natural androgens tested in transactivation assay, not just 11-Ketotestosterone and testosterone. Gen Comp Endocrinol, 117: 200–206.

Takeuchi, Y, Yoshizaki, G, Takeuchi, T. 2001. Production of germ-line chimeras by rainbow trout by blastomere transplantation. Mol Reprod Dev, 59: 380–389.

Takeuchi, Y, Yoshizaki, G, Kobayashi, T, et al. 2002. Mass isolation of primordial germ cells from transgenic rainbow trout carrying the green fluorescent protein gene driven by the vasa gene promoter. Biol Reprod, 67: 1087–1092.

Takeuchi, Y, Yoshizaki, G, Kobayashi, T, et al. 2003. Generation of live fry from intraperitoneally transplanted primordial germ cells in rainbow trout. Biol Reprod, 69: 1142–1149.

Takeuchi, Y, Yoshizaki, G, Kobayashi, T, et al. 2004. Surrogate broodstock produces salmonids. Nature, 430: 629–630.

Talbot, WS, Hopkins, N. 2000. Zebrafish mutations and functional analysis of the vertebrate genome. Genes Dev, 14: 755–762.

Tanaka, H, Hirose, K, Nogami, K, et al. 1990. Sexual maturation and sex reversal in red spotted grouper (*Epinephelus akaara*) populations. Bull Mar Sci, 52: 961–969.

Tanaka, K, Takehana, Y, Naruse, K, et al. 2007. Evidence for different origins of sex chromosomes in closely related *Oryzias* fishes: Substitution of the master sex determining gene. Genetics, 177: 2075–2081.

Tanaka, M, Saito, M, Morinaga, C, et al. 2008. Cross talk between germ cells and gonadal somatic cells is critical for sex differentiation of the gonads in teleost fish medaka (*Oryzias latipes*). Dev Growth Differ, 50: 273–278.

Taranger, GL, Carillo, M, Schulz, RW, et al. 2010. Control of puberty in farmed fish. Gen Comp Endocrinol, 165: 483–515.

Tatarenkov, A, Lina, SMQ, Taylor, DS, et al. 2009. Long term retention of self-fertilization in a fish clade. Proc Natl Acad Sci USA, 106: 14456–14459.

Tezuka, A, Yamamoto, H, Yokoyama, J, et al. 2011. The MC1R gene in the guppy (*Poecilia reticulata*): Genotypic and phenotypic polymorphisms. BMC Res Notes, 4: 31–38.

Thibault, RE, Schultz, RJ. 1978. Reproductive adaptations among viviparous fishes (Cyprinodontiformes: Poeciliidae). Evolution, 32: 320–333.

Thomas, P, Dressing, G, Pang, Y, et al. 2006. Progestin, estrogen, androgen G-protein coupled receptors in fish gonads. Steroids, 71: 310–316.

Trant, JM, Gavasso, S, Ackers, J, et al. 2001. Developmental expression of cytochrome p450 aromatase genes (*cyp19a* and *cyp19b*) in zebrafish fry (*Danio rerio*). J Exp Zool, 290: 475–483.

Tripathy, N, Hoffman, M, Willing, EM, et al. 2009. Genetic linkage map of the guppy *Poecilia reticulata* and quantitative loci analysis of male sex and colour vision. Proc R Soc London, 276B: 2195–2208.

Tse, AC-K, Ge, W. 2010. Spatial localization of EGF family ligands and receptors in the zebrafish ovarian follicle and their expression profiles during folliculogenesis. Gen Comp Endocrinol, 167: 397–407.

Turner, BJ, Elder, JF Jr, Laughlin, TF, et al. 1992. Extreme clonal diversity and divergence in populations of a selfing hermaphroditic fish. Proc Natl Acad Sci USA, 89: 10643–10647.

Turner, CL. 1941. Gonadal characteristics produced in the anal fin of female of *Gambusia affinis affinis* by treatment with ethinyltestosterone. Biol Bull, 80: 371–383.

Tyler, CR, Nagler, JJ, Pottinger, TG, et al. 1994. Effects of unilateral ovariectomy on recruitment and growth of follicles in the rainbow trout *Oncorhynchus mykiss*. Fish Physiol Biochem, 13: 309–316.

Tyler, CR, Pottinger, TG, Coward, K. 1997. Salmonid follicle-stimulating hormone (GtH-I) mediates vitellogenic development of oocytes in the rainbow trout *Oncorhynchus mykiss*. Biol Reprod, 57: 1238–1244.

Uchida, D, Yamashita, M, Kitano, T, et al. 2002. Oocyte apoptosis during the transition from ovary-like tissues to testes during sex differentiation of juvenile zebrafish. J Exp Zool, 205: 711–718.

Uhlenhaut, NH, Jakole, S, Anlag, K, et al. 2009. Somatic sex reprogramming of adult ovaries to testes by FoxL2 ablation. Cell, 139: 1130–1142.

Underwood, JL, Hestand, RSII, Thompson, BJ. 1986. Gonadal regeneration in grass carp following bilateral gonadectomy. Prog Fish Cult, 48: 54–56.

Valdesalici, S, Cellerino, A. 2003. Extremely short lifespan in the annual fish *Nothobranchius furzeri*. Proc R Soc London, 270B: S189–S191.

Valenzano, DR, Terzibasi, E, Genade, T, et al. 2006. Resveratol prolongs lifespan and retards the onset of age-related markers in a short-lived vertebrate. Curr Biol, 16: 296–300.

van den Hurk, R, Slof, GH. 1981. A morphological and experimental study gonadal sex differentiation in the rainbow trout *Salmo gairdneri*. Cell Tissue Res, 218: 487–497.

van der Sluijs, I, Sheehausen, D, Van Dooren, TJM, et al. 2010. No evidence for a genetic association between female mating preference and male secondary sexual trait in a Lake Victoria cichlid fish. Curr Biol, 56: 57–64.

van Tienhoven, A. 1968. Embryology. pp. 8–28. In: A van Tienhoven (ed) Reproductive Physiology of Vertebrates. WB Saunders, London, UK.
Vecsei, P, Litvek, MK, Noakes, DLG, et al. 2003. A non-invasive technique for determining sex of live North American sturgeons. Env Biol Fish, 68: 333–338 .
Veith, AM, Froschauer, A, Kortig, C, et al. 2003. Cloning of dmrt1 gene of *Xiphophorus maculatus*: dmy/dmrt1bY is not the master sex determining gene in the platyfish. Gene, 317: 59–66.
Veith, AM, Schafer, M, Kluver, N, et al. 2006. Tissue specific and adults of the platyfish *Xiphophorus maculatus*. Zebrafish, 3: 325–337.
Vizziano, R, Randuinean, G, Baron, D, et al. 2007. Characterization of early molecular sex differentiation in rainbow trout *Oncorhynchus mykiss*. Dev Dyn, 236: 2198–2206.
Volff, JN, Nanda, I, Schmid, M, et al. 2007. Governing sex determination in fish: Regulatory putsches and ephemeral dictators. Sex Dev, 1: 85–99.
von Hofsten, J, Olsson, PE. 2005. Zebrafish sex determination and differentiation: involvement of FTZ-F1 genes. Reprod Biol Endocrinol, 3: doi: 10.1186/1477-7827-3-63.
von Hofsten, J, Jones, I, Karsson, J, et al. 2001. Developmental expression patterns of FTZ-F1 homologoue in zebrafish (*Danio rerio*). Gen Comp Endocrimol, 121: 146–155.
von Hofsten, J, Larsson, A, Olsson, PE. 2005a. Novel steroidogenic factor 1 homolog (*ffd*) is co-expressed with anti-Mullerian hormone (*Amh*) relative to sox9a, sox9b and cyp19ala during gonad development. Gene Expr Patterns, 5: 655–667.
von Hofsten, J, Modig, C, Larsson, A, et al. 2005b. Determination of the expression pattern of the dual promoter of zebrafish *fushi tarazu* factor-1a following microinjections into zebrafish one cell stage embryo. Gen Comp Endocrinol, 142: 222–226.
Wakamatsu, Y, Ju, B, Pristyaznhyuk, I, et al. 2001. Fertile and diploid nuclear transplants derived from embryonic cells of a small laboratory fish medaka (*Oryzias latipes*). Proc Natl Acad Sci USA, 98: 1071–1076.
Wallace, BMN, Wallace, H. 2003. Synaptonemal complex karyotype of zebrafish. Heredity, 90: 136–140.
Walter, RB, Hazlewood, I, Kazianis, S. 2006. The *Xiphophorus* Genetic Stock Centre Manual. pp. 1–129. In: K Kallman, M Schartl (eds) Som Marcos. Texas State University Press, Texas, USA.
Wang, D-S, Zhou, L-Y, Kobayashi, T, et al. 2010. Doublesex-Mab3 related transcription factor-1 repression of aromatase transcription, a possible mechanism favouring the male pathway in tilapia. Endocrinology, 151: 1331–1340.
Wang, Li-H, Tsai, C-L, 2000. Effects of high temperatures on the deformity and sex differentiation of tilapia *Oreochormis mossambicus*. J Exp Zool, 286: 534–537.
Wang, XG, Orban, L. 2007. Anti-Mullerian hormone and 11β-hydroxylase show reciprocal expression to that of aromatase in the transforming gonad of zebrafish males. Dev Dyn, 236: 1329–1338.
Wang, XG, Bartfai, R, Sleptosova-Freidrich, I, et al. 2007. The timing and extent of juvenile ovary phase are highly variable during zebrafish testis differentiation. J Fish Biol, 70A: 33–44.
Wang, Y, Zhou, L, Yao, B, et al. 2004. Differential expression of thyroid-stimulating hormone beta subunit in gonads during sex reversal of orange-spotted and red-spotted groupers. Mol Cell Endocrinol, 220: 77–88.
Warner, RR. 1975. The adaptive significance of sequential hermaphroditism in animals. Am Natl, 182: 61–82.
Warner, RR. 1978. The evolution of hermaphroditism and unisexuality in aquatic and terrestrial vertebrates. In: ES Reese, FJ Lighter (eds) Contrasts in Behaviour. John Wiley & Sons, New York, USA.
Watanabe, M, Tanaka, M, Kobayashi, D, et al. 1999. Medaka (*Oryzias latipes*) FTZ-F1 potentially regulates the transcription of p450 aromatase in ovarian follicles: DNA cloning and functional characterization. Mol Cell Endocrinol, 149: 271–274.

Watanabe, T, Yoshida, M, Nakajima, M, et al. 2005. Linkage mapping of AFLP and microsatellite DNA markers with body colour and sex determining loci in the guppy (*Poecilia reticulata*). Zool Sci, 22: 883–892.

Wei, Q, Ke, F, Zhang, J. 1997. Biology, fisheries and conservation of sturgeons and paddlefish in China. Env Biol Fish, 48: 241–255.

Wen, A, You, F, Tan, X, et al. 2009. Expression pattern of *dmrt4* from olive flounder (*Paralichthys olivaceus*) in adult gonads and during embryogenesis. Fish Physiol Biochem, 35: 421–433.

Winkler, C, Hornung, U, Kondo, M, et al. 2004. Developmentally regulated and non-sex-specific expression of autosomal dmrt genes in embryos of the medaka fish (*Oryzias latipes*). Mech Dev 121: 997–1005.

Witthames, PR, Thorsen, A, Murus, H, et al. 2009. Advances in methods of determining fecundity: application of the new methods to some marine fishes. Fish Bull, 107: 148–164.

Wong, TT, Zohar, Y. 2004. Novel expression of gonadotropin subunit genes in oocytes of the gilthead seabream (*Sparus aurata*). Endocrinology, 145: 5210–5220.

Wong, TT, Fjiri, S, Zohar, Y. 2006. Molecular biology of ovarian aromatase in sex reversal: Complimentary DNA and 5 flanking region isolation and differential expression of ovarian aromatase in the gilthead seabream (*Sparus aurata*). Biol Reprod, 74: 857–864.

Wong, TT, Saito, T, Crodian, J, et al. 2010. Zebrafish germline chimeras produced by transplantation of ovarian germ cells into sterile host larvae. Biol Reprod, 84: 1190–1197.

Woodhead, AD. 1978. Fish in studies of aging. Exp Gerontol, 13: 125–140.

Woolcock, B, Kazianis, S, Lucito, R, et al. 2006. Allee specific marker generation and linkage mapping on the *Xiphophorus* sex chromosomes. Zebrafish, 3: 23–27.

Wooton, RJ. 1998. Ecology of Teleost Fishes. Kluwer Academic Publishers, Dordrecht, The Netherlands.

Woram, RA, Gharbi, K, Sakamoto, T, et al. 2003. Comparative genome analysis of the pimary sex determining locus in salmonid fishes. Genome Res, 13: 272–280.

Wu, G-C, Tomy, S, Chang, C-F. 2008. The expression of *nrob1* and *nr5a4* during gonad development and sex change in protandrous black porgy fish *Acanthopagrus schlegeli*. Biol Reprod, 78: 200–210.

Wu, G-C, Chiu, P-C, Lyu, Y-S, et al. 2010. The expression of *amh* and *amh2* is associated with the development of gonadal tissue and sex change in the protandrous black porgy *Acanthopagrus schlegeli*. Biol Reprod, 110.Doi: 10.1095/biolreprod.110.084681.

Wu, N, Yue, HM, Chen, B, et al. 2009. Histone H2A has a novel variant in fish oocytes. Biol Reprod, 81: 275–283.

Wu, SF, Zhang, H, Cairns, BR. 2011. Genes for embryo development are packaged in blocks of multivalent chromatin in zebrafish sperm. Genome Res, 21: 578–579.

Xia, W, Zhou, L, Yao, B, et al. 2007. Differential and spermatogenic cell specific expression of DMRT1 during sex reversal in protogynous hermaphroditic groupers. Mol Cell Endocrinol, 263: 153–172.

Xie, T, Spradling, A. 2000. A niche maintaining germ line stem cells in the *Drosophila* ovary. Science, 290: 328–330.

Xu, H, Gui, J, Hong, Y. 2005. Differential expression of *vasa* RNA and protein during spermatogenesis and oogenesis in the gibel carp (*Carassius auratus gibelio*), a bisexually and gynogenetically reproducing vertebrate. Dev Dyn, 233: 872–882.

Xu, HY, Li, MY, Gui, JF, et al. 2010. Fish Germ Cells. Sci China Life Sci, 53: 435–446.

Yamaha, E, Otani, S, Minami, A, et al. 2002. Dorso-ventral perturbation in goldfish embryos caused by heat- and pressure shock treatments for chromosome set manipulation. Fish Sci, 108: 313–319.

Yamaha, E, Murakami, M, Hada, K, et al. 2003. Recovery of fertility in male hybrids of a cross between goldfish and common carp by transplantation of PGC (Primordial germ cell) containing graft. Genetica, 119: 121–131.

Yamaguchi, A, Lee, KH, Fujimoto, H, et al. 2006. Expression of the DMRT gene and its role in early gonadal development of Japanese pufferfish, *Takifugu rubripes*. Comp Biochem Physiol, D1: 59–68.

Yamaguchi, T, Yamaguchi, S, Hirai, T, et al. 2007. Follicle-stimulating hormone signaling and Foxl2 involved transcriptional regulation of aromatase gene during gonadal sex differentiation in Japanese flounder *Paralichthys olivaceus*. Biochem Biophys Res Commun, 359: 935–940.

Yamamoto. T. 1969. Sex differentiation. pp. 117–175. In: WS Hoar, DJ Randall (eds) Fish physiology, Academic Press, New York, USA.

Yano, A, Nicol, BJ, Guerin, A, et al. 2011. The duplicated rainbow trout (*Oncorhynchus mykiss*) T-Box transcription factors 1 *tbxla* and *tbx1b* are up-regulated during testicular differentiation. Mol Reprod Dev, 78: 172–180.

Yao, B, Zhou, L, Giu, JF. 2003. Studies on cDNA cloning and temporal and spatial expression of sox3 gene in grouper *Epenephelus tauvina*. High Technol Lett, 13: 74–81.

Yao, B, Zhou, L, Wang, Y, et al. 2007. Differential expression and dynamic changes of sox3 during gametogenesis and sex reversal in protogynous hermaphroditic fish. J Exp Zool, 307A: 207–219.

Yasui, GS, Fujimato, T, Sakao, S, et al. 2011. Production of loach (*Misgurnus anguillicaudatus*) germ line chimera using transplantation of germ cells isolated from cryopreserved blastomeres. J Anim Sci, 89: 2380–2388.

Yokoi, H, Kobayashi, T, Tanaka, M, et al. 2002. Sox9 in teleost fish medaka (*Oryzias latipes*): evidence for diversified function of Sox9 in gonad differentiation. Mol Reprod Dev, 63: 5–16.

Yoon, C, Kawakami, K, Hopkins, N. 1997. Zebrafish *vasa* homologue RNA is localized to the cleavage planes of 2 and 4-cell stage embryos and expressed in primordial germ cells. Development, 124: 3157–3166.

Yoshinaga, N, Shiraishi, E, Yamamoto, T, et al. 2004. Sexually dimorphic expression of a teleost homologue of Mullerian inhibiting substance during gonadal sex differentiation in Japanese flounder *Paralichthys olivaceus*. Biochem Biophy Res Commun, 322: 508–513.

Yoshizaki, G, Takeuehi, Y, Sakatani, S, et al. 2000. Germ cell specific expression of green fluorescent protein in transgenic rainbow trout under control of the rainbow trout *vasa*-like gene promoter. Int J Dev Biol, 44: 323–326.

Yoshizaki, G, Takeuehi, Y, Kobayashi, T, et al. 2002. Primordial germ cells: the blueprint for a piscine life. Fish Physiol Biochem, 26: 3–12.

Yoshizaki, G, Ichikawa, M, Hayashi, M, et al. 2010. Sexual plasticity of ovarian germ cells in rainbow trout. Development, 137: 1227–1230.

Yoshizaki, G, Fuyinuma, K, Iwasaki, Y, et al. 2011. Spermatogonial transplantation in fish: A novel method for the preservation of genetic resources. Comp Biochem Physiol, Part D, 6: 55–61.

Ytteborg, E, Baeverfjord, G, Torgersen, J, et al. 2010. Molecular pathology of vertebral deformities in hypothermic Atlantic salmon (*Salmo salar*). BMC Physiol, 10: 12.

Yu, L, Tucci, V, Kishi, S, et al. 2006. Cognitive aging in zebrafish. PLoS ONE, DOI:10.137/journal.pone.0000014.

Zanaria, E, Muscatelli, F, Bardoni, B, et al. 1994. An unusual number of the nuclear hormone receptor superfamily responsible for sex linked adrenal hypoplasia congenital. Nature, 372: 631–641.

Zarkower, D. 2006. Somatic sex determination. In: WormBook, The *C. elegans* Research Community (eds), WormBook, doi/10.1895/wormbook.184.1.

Zbikowska, HM. 2003. Fish can be first-advances in fish transgenesis for commercial applications. Transgenic Res, 12: 379–389.

Zhang, D, Xiong H, Mennigen, JA, et al. 2009. Defining global neuroendocrine gene expression patterns associated with reproductive seasonality in fish. PLoS ONE, 4: e5816.

Zhang, J. 2004. Evolution of *Dmy* a newly emergent male sex determining gene of medaka fish. Genetics, 166: 1887–1895.

Zhang, L, Zhu, T, Lin, D, et al. 2010. A second form of *sox11* homologue identified in the orange-spotted grouper (*Epinephelus coioides*): Analysis of sequence and mRNA expression pattern. Comp Biochem Physiol, 157B: 415–422.
Zhang, Q, Nakayama, I, Fujiwara, A, et al. 2001. Sex identification by male specific growth hormone pseudogene (GH-ψ) in *Oncorhynchus masou* complex and a related hybrid. Genetica, 111: 111–118.
Zhang, W, Zhang, Y, Zhang, L, et al. 2007. The mRNA expression of p450 aromatase, gonadotropinβ subunit and FTZ-F1 in the orange-spotted grouper (*Epinephelus coioides*) during 17α-methyltestosterme-induced precocious sex change. Mol Reprod Dev, 74: 665–673.
Zhang, XY, 2005. Primary studies on sex reversal factors possibly involved in the gonadal sex determination and differentiation of Southern catfish *Silurus meridionalis*. PhD Thesis, Southwest Normal Univ, China.
Zhang, X, Zha, J, Wang, Z. 2008. Influences of 4 nonylphenol on *doublesex-and mab-3-related transcription factor 1* gene expression and vitellogenin mRNA induction of rare minnow (*Gobiocypris rarus*). Environ Toxicol Chem, 27: 196–205.
Zhang, Z, Wang, Y, Wang, S, et al. 2011. Transcriptome analysis of female and male *Xiphophorus maculatus* Jp 163A. PLoS ONE, 6: e18379. doi: 10.1371/journal. pone. 0018379.
Zhao, R, Xuan, Y, Li, X, et al. 2008. Age related changes of germline stem cell activity, niche signaling activity and egg production in *Drosophila*. Aging Cell, 7: 344–354.
Zhou, L, Gui, J-F. 2008. Molecular mechanisms underlying sex change in hermaphroditic groupers. Fish Physiol Biochem, DOI 10.1007/s 10695-008-9219-0.
Zhou, R, Chang, H, Tiersch, TR. 2002. Differential genome duplication and fish diversity. Rev Fish Biol Fish, 11: 331–351.
Zimmerer, EJ, Kallman, KD. 1989. The genetic basis for alternative reproductive tactics in the pygmy swordtail, *Xiphophorus nigrensis*. Evolution, 43: 1298–1307.
Zupanc, GKH. 1985. Fish and their Behaviour. How Fishes Live: Specially written for Aquarists. Tetra Press, Melle, Germany.

Author Index

Species Index

Subject Index

M

N

O

P

Q

R

Color Plate Section

Chapter 1

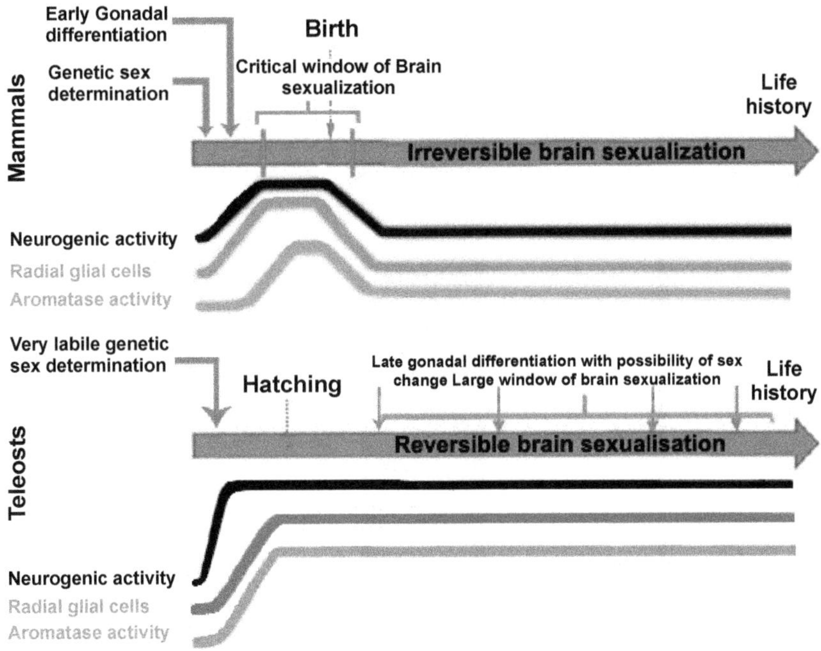

Fig 1. Ontogenetic pathways of sex determination and differentiation in mammals and fishes. Note the differences in the critical window of brain sexualization (adopted from Le Page et al., 2010)

Chapter 2

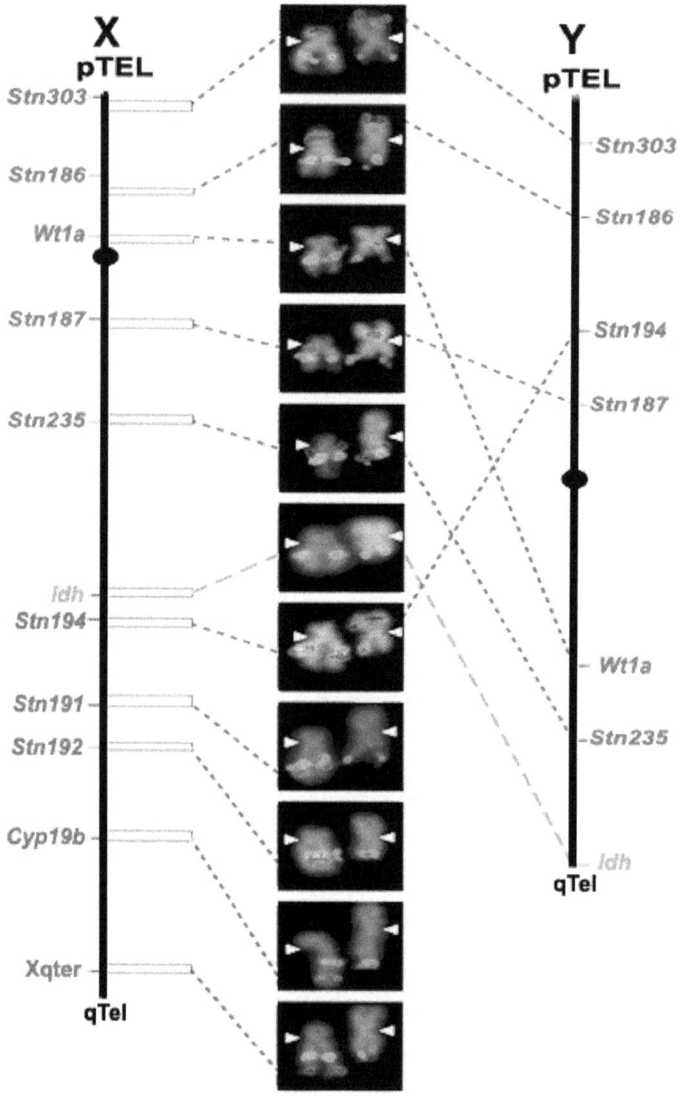

Fig. 5. FISH-based cytogenetic maps of X (left) and Y (right) chromosomes of *Gasterosteus aculeatus*. Each FISH section (center) contains the X and Y from a single metaphase spread, to which *Idh* and an additional probe are hybridized. The position of the centromere is identified with a white arrowhead. Dashed lines joining the X map to the FISH section indicate the BHC probe used in each section and those to the Y cytogenic map indicate the physical interval to which each FISH probe hybridizes. The black circles represent the positions of the centromeres. Note the deletion of *cyp19b* in the Y (from Ross and Peichel. 2008, with permission by Genetic Society of America)

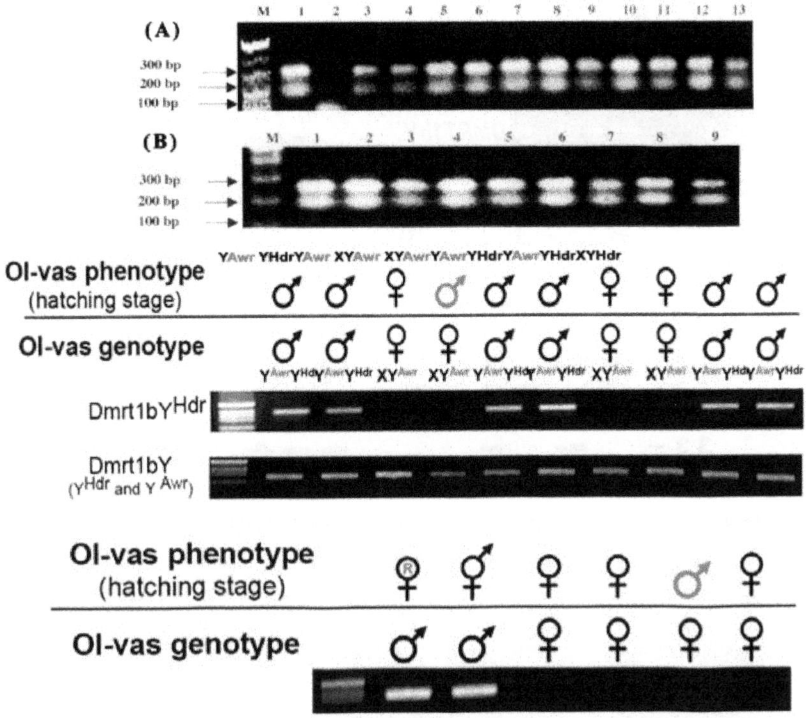

Fig. 6. Upper panel: PCR products of the genomic DNA of *Hemigrammus caudovittatus* as amplified by Dmrt1 primer. A row: Lane 1 = male, Lane 2 = female, Lanes 3–13 androgenic males generated using fresh or cadaveric sperm. B row: Lanes 1–7 unexpected female progenies and Lanes 8 and 9 male progenies (from David, 2004). Lower panel: *Oryzias latipes* progenies sired by an Ol-vas XY^{Hdr} ♂ crossed with $Y^{Awr}Y^{Awr}$ female. Note the absence of *Dmrt1bY*^{Hdr} expression on genotypic and phenotypic females and also the absence of expression of Ol-vas in genotypic and phenotypic females but its presence in males including a phenotypic hermaphrodite (from Herpin et al., 2007)

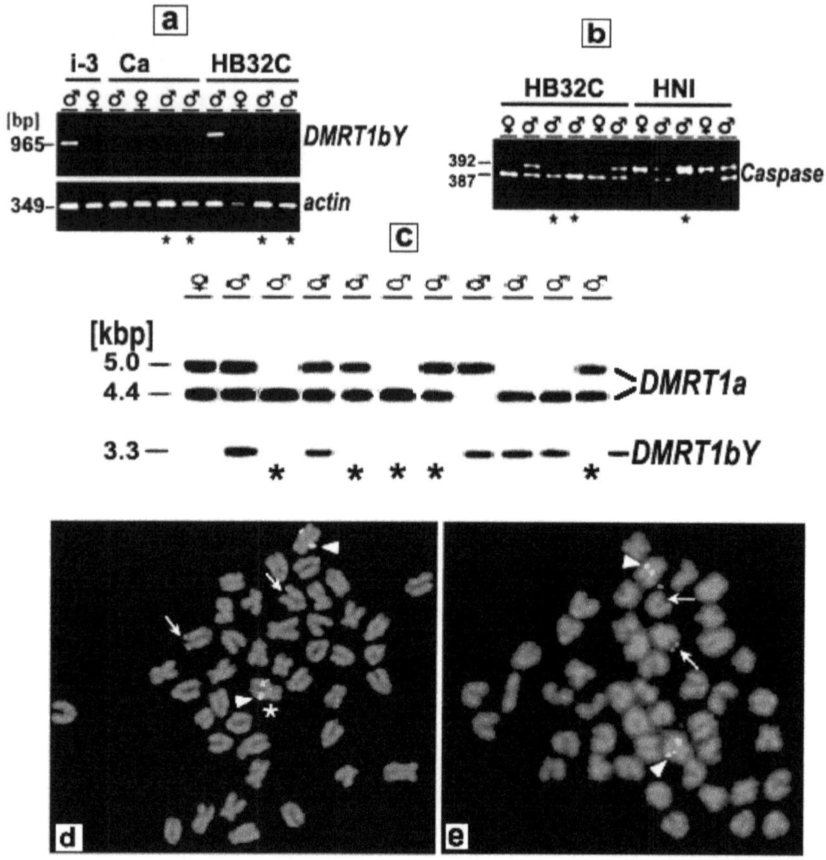

Fig. 10. *Oryzias latipes*: (A): PCR genotyping of males and females. Note the presence of *Dmrt1bY* in normal males but its absence in aberrant females. (B): Hemizygosity of normal males of the HB32C and HNI strains for *caspase* 6. Females and aberrant males (*) show only the X chromosomal PCR product. (C): Southern blot analysis of DNA from female, normal male and aberrant (*Dmrt1bY* negative) males (*). FISH pattern of Y specific and sex chromosome specific probes on (D) XY and (E) XX metaphase chromosomes. Note the presence of three hybridization signals in XY males, as compared to two spots in XX males (from Nanda et al., 2003, with permission by the Genetic Society of America).

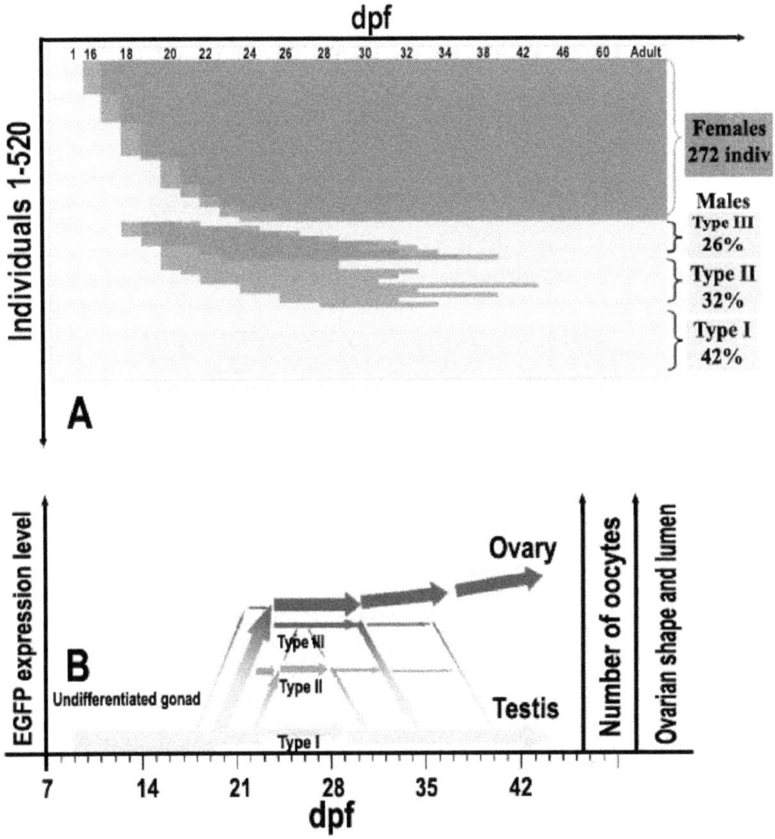

Fig. 15. *Danio rerio:* (A) Upper panel: The zygotic *vas::egfp* commences to highly express after 16 dpf in females but males of types I, II and III, which began to faintly express, progressively decrease the expression at different dpf (day post-fertilization). The increasing fluorescence intensity suggests the increasing oocyte number and growth of ovarian lumen and shape. In contrast the decreasing fluorescence indicates the degeneration of oocytes and transformation of ovary into testis. (B) Lower panel: A model for gonad differentiation showing high levels of variation in the intensity, onset and duration of gonad transformation (from Wang et al 2007. The timing and extent of the juvenile ovary phase are highly variable during zebrafish testis differentiation. J Fish Biol, 70A: 33–44, reproduced by permission of The Fisheries Society of the British Isles/John Wiley & Sons Ltd)

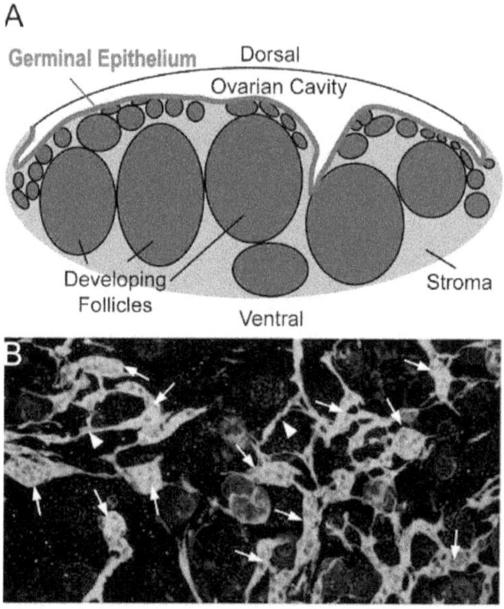

Fig. 33. (A) Transverse section of medaka ovary consisting of (i) ovarian and (ii) stromal compartments. Between these two compartments, note the presence of germinal epithelium lined by the basement membrane. (B) A dorsal view of the ovarian cord in ovary of *sox9b*-EGFP transgenic medaka. 'Germinal cradles' (arrows) composed of germ cells and the *sox9b*-expressing cells are connected to each other by the cellular processes of *sox9b*-expressing cells (arrow heads) to form the network, the 'ovarian cords' (from Nakamura et al., 2011. Ovarian germline stem cells in the teleost fish, medaka (*Oryzias latipes*). Int J Biol Sci, 7: 403–409, with permission of IJBS Publishing Team)

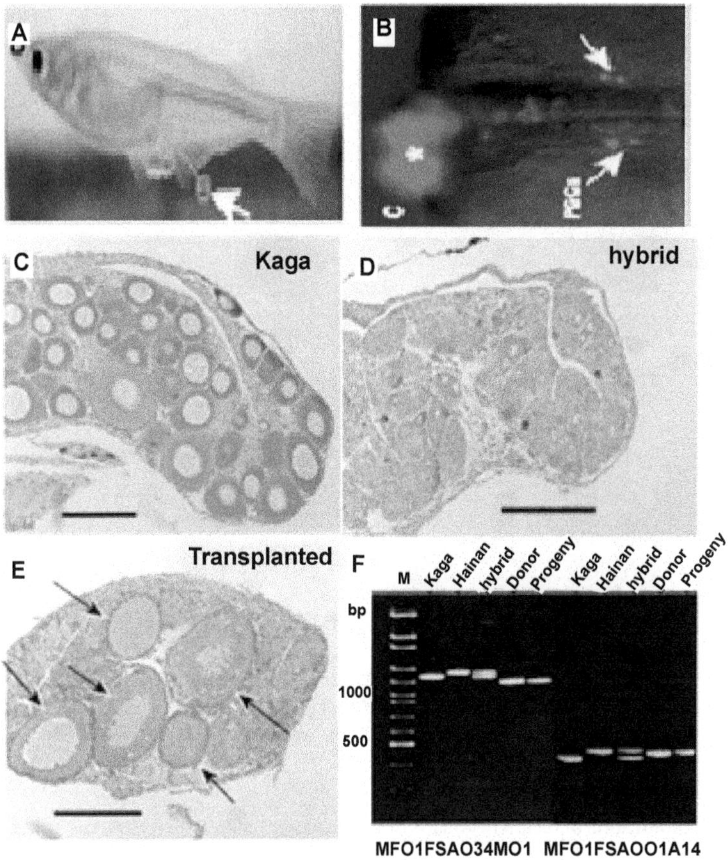

Fig. 34. Recovery of fertility in sterile hybrid medaka by PGCs transplantation from fertile medaka. A = medaka with arrow indicating egg being spawned, B = colonization of 'undifferentiated' sterile gonad by transplanted PGCs, C = Cross sections of ovaries from a 4 week old Kaga strain , D = Non transplanted hybrid, E =Transplanted hybrid ovary contains growing oocytes, as shown by arrows. Scale bar, 200 μm and F = Confirmation of germ-line transfer by detection of strain-specific genetic markers. Note that the patterns of marker amplification for the progeny are identical to those of the donor (from Shimada and Takeda, 2008)

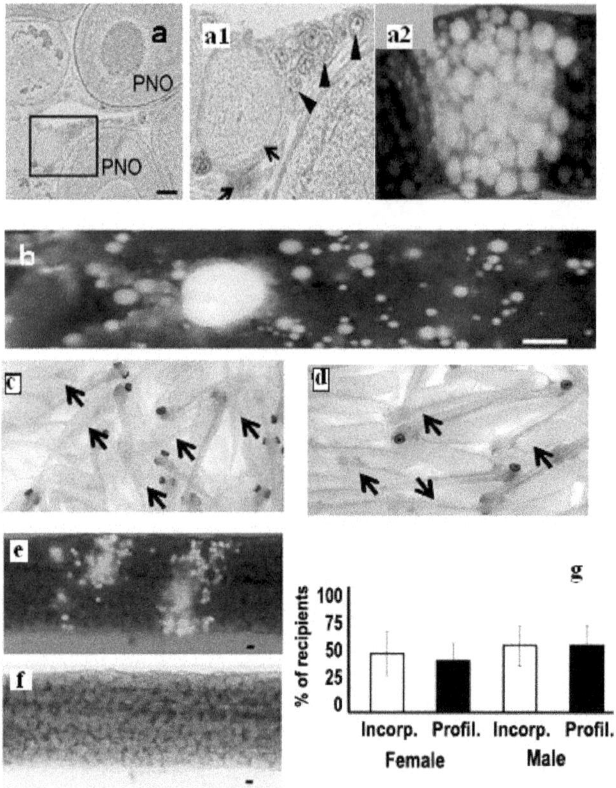

Fig. 35. Colonization of recipient gonads by ovarian germ cells of *Oncorhynchus mykiss*. (A-A_1) perinuclear oocytes (PNO) and oocytes (arrow heads) showing *Gfp*—specific antibody. (A2) *Gfp*-labeled donor germ cells proliferated and differentiated into oocytes in recipient ovary. (B) *Gfp*–labeled donor-derived germ cells remaining in the post-ovulatory follicles of 3n recipient. 50% F_1 progenies were OSCs-derived orange colored females (C) and males (D). Fluorescent (E) and bright-field (F) views of colonization and proliferation of *Gfp*-labeled donor germ cells in the testes of 5 month old recipients. (g) The percentage of progenies carrying the donor-derived germ cells in their gonads (black bars) and those, who did not differ significantly between female recipients and male recipients (from Yoshizaki et al., 2010. Sexual plasticity of ovarian germ cells in rainbow trout. Development, 137 : 1227–1230, doi : 10.1242/dev 044982, reproduced with kind permission by The Company of Biologists Ltd)

Chapter 4

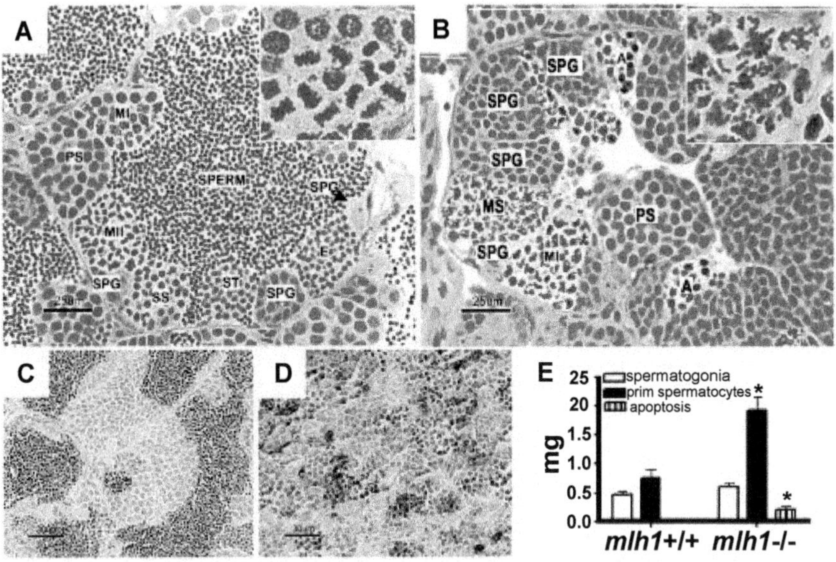

Fig. 60. Cross sections of seminiferous tubule in (A) *mlh1*$^{+/+}$, (B) *mlh1*$^{-/-}$ zebrafish, showing spermatogenic cysts with different types of germ cells, spermatogonia (SPG), primary spermatocytes (PS), first meiotic division (MI), secondary spermatocytes (SS), second meiotic division (MII), spermatids (ST), spermatozoa (SPERM), and apoptotic spermatocytes (A). Note the absence of postmeiosis I stages and the presence of apoptotic spermatocytes with strongly condensed nuclei in *mlh1*$^{-/-}$ testis. Inset shows a magnified view of the first meiotic division. In the wild type, chromosomes are aligned at the cell equator just prior to division. In the mutant, chromosomes are randomly distributed throughout the nucleus. TUNEL staining of (C) *mlh1*$^{+/+}$ and (D)*mlh1*$^{-/-}$. High numbers of apoptotic spermatocytes are seen in mutant testis. Wild-type testis shows a very low incidence of apoptosis. (E) Morphometric analysis of testes sections showing increase in amounts of spermatocytes and apoptoic cells, but not of spermatogonia. (from Feitsma et al., 2007, with permission by the Genetic Society of America)

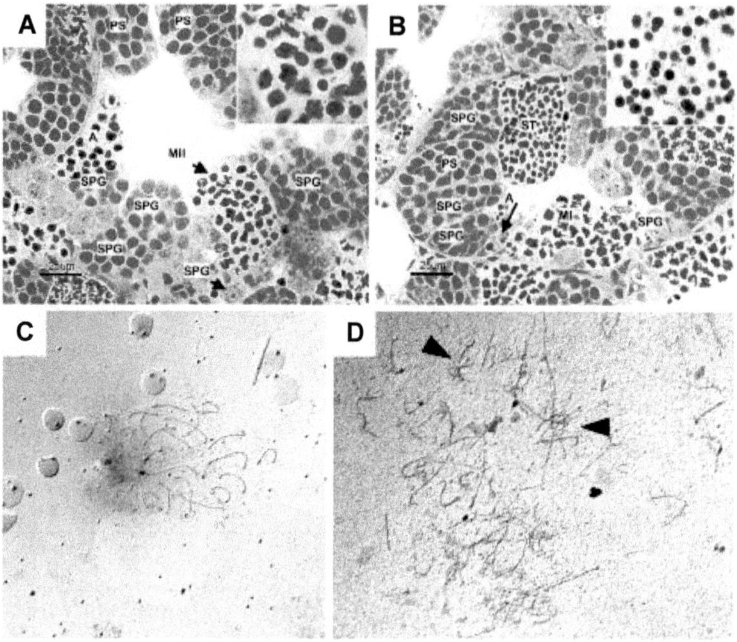

Fig. 61. Cross sections of seminiferous tubules in (A) infertile and fertile (B) triploid zebrafish. For expansion of abbreviations, see Fig. 4.3. Inset in (A): magnification of second meiotic division; note the concomitant presence of secondary spermatocytes and spermatids. (F) and G). Synapsis in triploid spermatocytes (D) is strongly disorganized as compared to that in wild type (C). Irregular synapsis is indicated by arrow heads (from Feitsma et al., 2007, with permission by the Genetic Society of America)

For Product Safety Concerns and Information please contact our EU representative GPSR@taylorandfrancis.com
Taylor & Francis Verlag GmbH, Kaufingerstraße 24, 80331 München, Germany

www.ingramcontent.com/pod-product-compliance
Ingram Content Group UK Ltd.
Pitfield, Milton Keynes, MK11 3LW, UK
UKHW020201200926
475702UK00004B/555
9781578087990